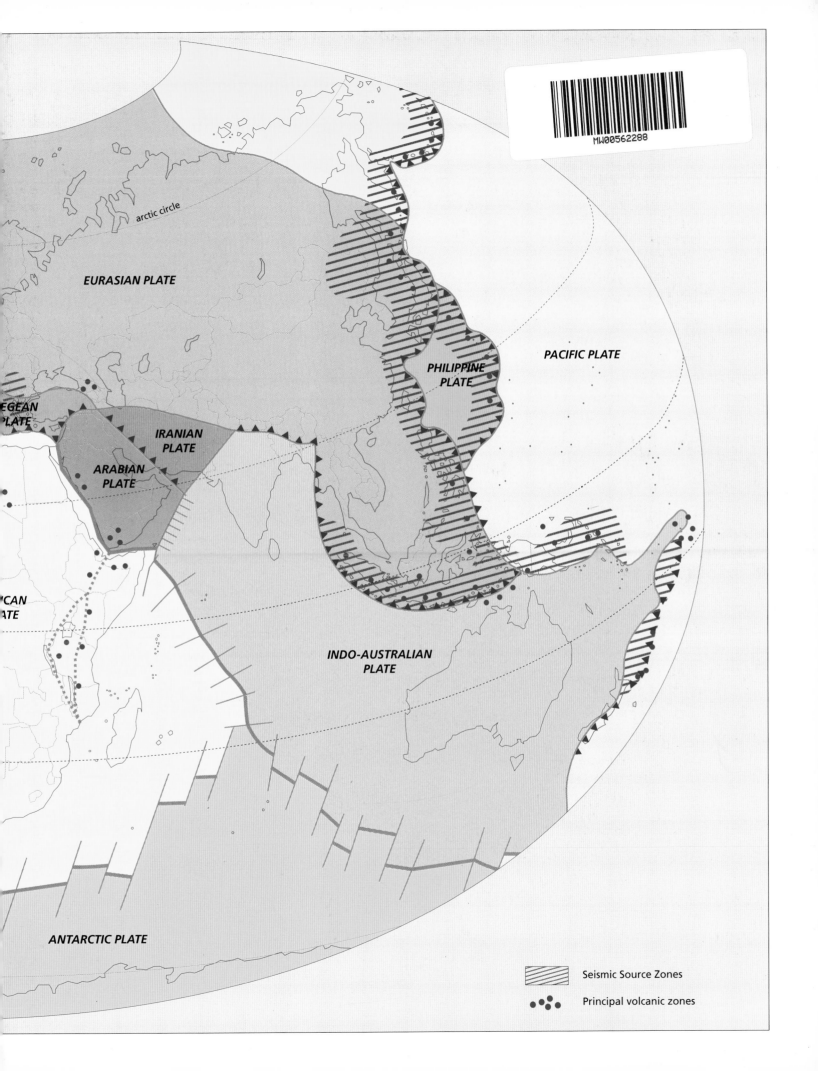

EURASIAN PLATE

arctic circle

PACIFIC PLATE

PHILIPPINE
PLATE

IRANIAN
PLATE

EGEAN
LATE

ARABIAN
PLATE

'CAN
ATE

INDO-AUSTRALIAN
PLATE

ANTARCTIC PLATE

MW00562288

▨ Seismic Source Zones

•••• Principal volcanic zones

Great Wine Terroirs

University of California Press
Berkeley and Los Angeles, California

University of California Press, Ltd.
London, England

Originally published in French by Hachette Livre (Hachette Pratique), 2001.

Library of Congress Cataloging-in-Publication Data

Fanet, Jacques.
    Great wine terroirs / Jacques Fanet.
        p. cm.
    Includes bibliographical references and index.
    ISBN 0–520–23858–3 (cloth : alk. paper)
        1. Wine and wine making—France—Regional disparities.   I. Title.

TP553.F35 2004
641.2'2—dc22

                                                                2004047958

Manufactured in China

13 12 11 10 09 08 07 06 05 04

10 9 8 7 6 5 4 3 2 1

Great Wine Terroirs

By Jacques Fanet

Translated from the French

by Florence Brutton

UNIVERSITY OF CALIFORNIA PRESS

Berkeley   Los Angeles   London

Geological Map of France

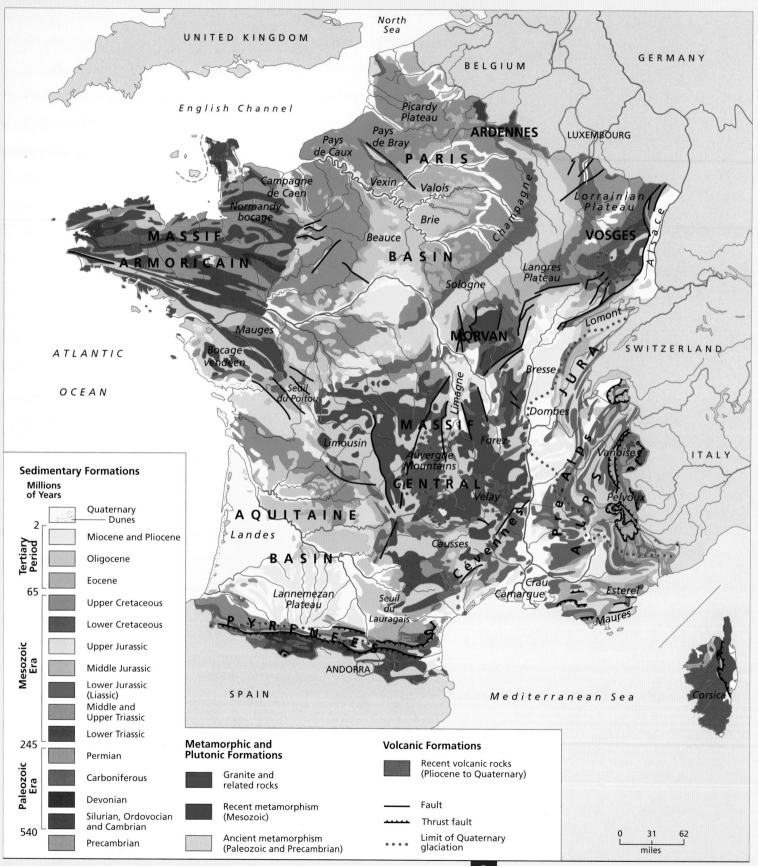

North
Sea

UNITED KINGDOM

GERMANY

BELGIUM

English Channel

Picardy
Plateau

LUXEMBOURG

ARDENNES

Pays
de Bray

Pays
de Caux

PARIS

Vexin

Valois

Lorrainian
Plateau

Campagne
de Caen

Champagne

VOSGES

Normandy
bocage

Brie

MASSIF

Beauce

BASIN

Alsace

ARMORICAIN

Langres
Plateau

Sologne

Lomont

Mauges

MORVAN

SWITZERLAND

ATLANTIC

Bocage
vendéen

Bresse

JURA

OCEAN

Seuil
du Poitou

Dombes

Limagne

MASSIF

Vanoise

ITALY

Limousin

Forez

Pre-ALPS

ALPS

Auvergne
Mountains

Pelvoux

CENTRAL

Velay

AQUITAINE

Landes

Causses

Cévennes

Esterel

BASIN

Crau
Camargue

Maures

Lannemezan
Plateau

Seuil
du
Lauragais

P Y R E N E E S

ANDORRA

SPAIN

Mediterranean Sea

Corsica

Sedimentary Formations

**Millions
of Years**

| | Quaternary Dunes |

Tertiary Period

2

Miocene and Pliocene

Oligocene

Eocene

65

Mesozoic Era

Upper Cretaceous

Lower Cretaceous

Upper Jurassic

Middle Jurassic

Lower Jurassic
(Liassic)

Middle and
Upper Triassic

Lower Triassic

245

Paleozoic Era

Permian

Carboniferous

Devonian

Silurian, Ordovician
and Cambrian

540

Precambrian

Metamorphic and Plutonic Formations

Granite and
related rocks

Recent metamorphism
(Mesozoic)

Ancient metamorphism
(Paleozoic and Precambrian)

Volcanic Formations

Recent volcanic rocks
(Pliocene to Quaternary)

⸺ Fault

⊢⊣⊣⊣ Thrust fault

•••• Limit of Quaternary
glaciation

0   31   62

miles

8

Viticultural Map of France

In addition to flourishing vineyards right along the Mediterranean coastline (in the former Roman province of Narbonne), vines are also planted along the major rivers (the Garonne, Loire, Rhône and their tributaries) and on sheltered slopes protected from north winds (in Alsace, Burgundy, Champagne and the northern Rhône Valley).

Terroir: Myth or Reality?

At the dawn of the third millennium, there is what amounts to a positive food revolution in Europe as country after country abandons an agricultural policy that has been in force for more than 50 years. After the years of stagnation then shortages that marked the first half of the twentieth century, agricultural research backed by European governments and producers focused on making Europe self-sufficient in food production. What followed were new intensive farming methods which treated soil as an inert substrate that could simply be "laced" with synthetic fertilizers and water to improve productivity. While the policy did help to feed Western populations it also led to agricultural surpluses and growing consumer demand for increased food safety and traceability. Consumers these days want to know what they are eating, where it comes from and how it is produced. After 50 years of standardized food products, they also want an increasingly varied diet.

Enhancing the Natural Environment

In the same period, another system of agricultural production was steadily winning converts in southern Europe and especially in France, based on the concept of Appellations d'Origine Contrôlée (Controlled Origin Appellations). This system rested on a principle that was diametrically opposed to intensive farming: limited production, governed by hard and fast regulations. Its raison d'être was to enhance the natural environment (soil, climate) using sustainable, traditional methods that lead to distinctive products with characteristics so closely linked to the place of harvest that they could not be produced anywhere else.

Far from being out-of-date, as some have said in the past, the Appellations d'Origine system has never been more relevant. Indeed, following its highly successful application in the wine area, it has been adopted by a wide range of other agricultural sectors. What holds it all together is that uniquely French notion of *terroir*: an umbrella term for a subtle interaction of natural factors and human skills that define the characteristics of each wine-growing area. There is no equivalent term in any other language, which may explain why it is occasionally misapplied. New World producers for their part strongly dispute any link between a wine and the soil that produces it. They say there is not a shred of evidence for such a claim and that the whole idea was dreamed up by French producers to keep out foreign competition. By restricting production to so-called reference areas, the AOC concept makes a wine more exclusive and more expensive as a result. You can see why New World producers think the way they do. When vines were first planted in the Southern Hemisphere, the rootstock was European and so was the know-how. The only thing colonial planters could not bring with them from the Old World were the soils. The settlers'

first priority in any case was to find climate conditions compatible with vine cultivation rather than worry about the type of soil.

What's more, viticultural practice in the Old World—choice of sites, growing techniques and appropriate vines—is based on a trial-and-error process dating back at least two thousandyears to the start of wine-growing in Europe. New World viticulture, by comparison, is still in its infancy. It officially started 400 years ago but only really got going a century and a half ago and much more recently in some countries. So it is hardly surprising if the link between wine and soil remains largely unknown.

Discovering a Subtle Link

In fairness to people who dispute the notion of terroir, it has to be said that demonstrating its influence is not easy and depends on certain conditions. It is no accident that the concept has been championed by winegrowers in Burgundy (eastern France), a patchwork of vineyards where each grower tends to make his own wine in his own way.

If Burgundy wines are to express the subtle variations of the soil, grapes from each separate parcel (vineyard site) must be vinified individually. Furthermore, the person who grows and harvests the grapes must also supervise their conversion into wine, tasting the wines regularly to compare them as they develop. No one region better illustrates the notion of the terroir than Burgundy. Witness the wealth of AOC Burgundy wines, the characteristic small Burgundy-style barrels and the tradition of making wines from one red grape (Pinot Noir) and one white grape (Chardonnay). The multiplicity of terroirs in Burgundy, known locally as *climats*, was first recognized and exploited by the monks in the twelfth century.

Elsewhere in France it is more difficult, but by no means impossible, to demonstrate the impact of the soil on the wine, since the practice is to blend together different varietals from different sites.

The Nature of the Soil-Wine Link

There are two main forces at work in the soil-wine relationship: Nature and people. Nature provides the raw material, in this case the soil, but only people know how to turn it into something exceptional. All the great French Grands Crus, from the gravel mounds (*Graves*) of the Médoc, the steam-rolled cobbles of Châteauneuf-du-Pape, or the Jurassic limestone slopes of the Côte de Nuits, owe their renown to generations of winegrowers.

The soil in a vineyard may be compared to the film in a camera. The film contains a record of all the images shot by the photographer but is itself of no interest. Open up the camera and all you will see is a blank strip of vinyl. You need developers, fixers, and photographic paper to reveal those invisible images. In the same way, you need appropriate varietals and skills to produce wines that fully express the many nuances of the terroir.

Time-Honored Creation

Why does one particular type of soil produce one particular type of wine? Curiously, there are very few in-depths studies of the role played by the soil, perhaps because of the cost and the many factors involved. The only published findings are based on two studies conducted by researchers in France, both of which provide incontrovertible scientific evidence of a link between wine and soil.

The first study, conducted by Professor Gérard Seguin and his colleagues from the University of Bordeaux (in southwest France), highlighted the role of water in relation to the quality of a cru. They demonstrated that the soils underlying most of the Bordeaux Grands Crus provide the vines with a perfectly homogeneous

water supply throughout their growth cycle—-so debunking the age-old theory that vines have to suffer to produce great wines. A certain lack of water does lead to more complex wines by restricting leaf growth and causing the sugars produced through photosynthesis to be deposited in the berries. Extreme water shortages on the other hand put the vines under stress and stop the berries from ripening. Professor Seguin and his team concluded that soil structure, permeability, and speed of water flow all play a major role in vineyard quality.

More recently, researchers led by René Morlat and Christian Asselin from the French "Institut National de la Recherche Agronomique" (INRA: National Institute of Agronomical Research) confirmed the importance of a stable water supply and also demonstrated the importance of soil temperature and the soil's ability to warm up quickly. Their findings were based on a 15-year observation period in three famous vineyards of the Touraine (northwest central France): Saumur-Champigny, Bourgueil, and Chinon. There have been no other studies to date, although the link between wine and its place of origin is beginning to arouse interest in new wine-producing countries. California producers, for example, are involved in a program of research led by the Department of Viticulture and Enology at the University of California, Davis. Market forces no doubt have something to do with it. Faced with a global demand for varietal wines but no more than ten international varieties at their disposal, producers have to distinguish themselves from their competitors. One way is to keep prices down; another is to develop new brands and another, chosen by growing numbers of producers, is to identify with a specific vineyard and methods of production. New World wines increasingly display place names, although not all of these names suggest any particular characteristics.

However, we are also seeing wines that correspond more closely to the European notion of controlled origin—-wines whose originality is plainly determined by the soil or the climate. Examples include Chilean Casablanca Chardonnay, Argentine Torrontes Famatina Valley, and Australian Coonawarra Cabernet Sauvignon. There is also the Stags Leap Wine Cellars, in California's Napa Valley, that vinifies grapes from each parcel separately.

In the following pages we review the principal wine-growing terroirs of France, the rest of Europe and the New World, covering both pedology (the study of soils) and geology. Note that soil study cannot replace geology because in many areas, especially in sloping vineyards, the soil is not derived exclusively from the parent rock but rather from a subtle combination of elements found in the stratum outcropping along the slope. The explanations provided here should answer the questions most frequently asked by wine lovers. Why, for example, is the Châteauneuf plateau, almost 400 feet above the Rhône Valley, surrounded by river alluvia? Why are there fossilized oysters in the soils of Chablis? As the book shows, each era in the geological timescale had a specific impact on the development of great vineyards and the characteristics of wine-growing regions may be classified according to their geological origin.

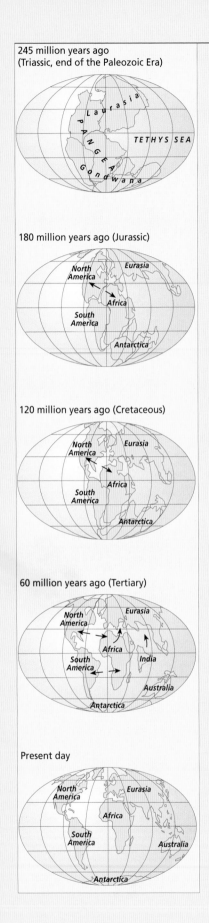

245 million years ago
(Triassic, end of the Paleozoic Era)

Laurasia
PANGEA
TETHYS SEA
Gondwana

180 million years ago (Jurassic)

North America
Eurasia
Africa
South America
Antarctica

120 million years ago (Cretaceous)

North America
Eurasia
Africa
South America
Antarctica

60 million years ago (Tertiary)

North America
Eurasia
Africa
India
South America
Australia
Antarctica

Present day

North America
Eurasia
Africa
South America
Australia
Antarctica

Continental Drift and the Origins of Terroirs

It seems implausible that the geographical development of the vineyards could have been affected by a shift in the positions of the Earth's continents that started 200 million years ago. At one end of the geological timescale there are landscapes shaped by men; at the other, there are monumental upheavals so slow that they defy the imagination. In fact, the succession of landslides, collisions, eruptions, erosions, and subductions that marked the Paleozoic and Mesozoic Eras and the Tertiary Period was indeed the original source of the rocks and great masses that were then chiseled into shape 2 million years ago, at the start of the Quaternary Period, to form the wine-growing landscapes that we see today.

In 1915, Alfred Wegener, the German geologist, meteorologist and Arctic explorer, published his theory of continental drift. He suggested that the Earth's continents moved gradually over the surface of the planet on a substratum of magma. Wegener pointed to the near perfect fit between the Americas and the land masses of Europe and Africa, and to the identical mineral deposits, rocks, and animal fossils found on opposite sides of the Atlantic Ocean. He suggested that the continents on either side of the ocean were slowing drifting apart. Wegener's theory was dismissed as fantasy by the leading specialists of the time and would not be revived until the 1960s, when a number of key discoveries shed new light on the subject. In particular, studies of the ocean floor had revealed long undersea ranges, called ocean ridges, in the middle of the Atlantic, equidistant from both shores.

When Economic Crises Bring Geology into Focus

What, then, is the link between the soil in great vineyards and continental drift? It is in fact a direct, fundamental link that became glaringly obvious in France following the phylloxera epidemic. By the mid-nineteenth century, French vineyards were no longer confined to the south of France (Provence and the Languedoc), as they had been in Roman times, but extended as far north as the suburbs of Paris, and even into Normandy. In the late twentieth century, the devastation wreaked by the phylloxera louse had reduced the vast French vineyard to a third of its former size. Wine growing remains significant in Mediterranean regions but survives in the west and northwest only in privileged locations where the relief and geology favor viticulture despite often adverse weather conditions.

Many of these regions, now mostly classified as AOC, owe their existence to one major geological phenomenon: the collision of the European and African continents that uplifted the Alps and the Pyrenees, causing varying degrees of crustal upheaval throughout the French territory.

A Speeded-up Film

Let us now take an accelerated view of a process that was at work for more than 100 million years. Imagine the Earth at the end of the Paleozoic Era, 250 million years ago. Most of its surface was below sea level except for a few areas that were grouped together in one huge landmass which geologists call *Pangaea*. In fact, the Precambrian Era (lasting more than three million years) and the Paleozoic Era were periods of intense orogeny, or mountain building, caused by collisions of varying intensity between the plates at the surface of the Earth. In the Hercynian, the final orogeny, all of these ridges and valleys came together to form one single mountain chain. In the Triassic Period of the Mesozoic Era, this range split into two: a northern part, known as *Laurentia*, that would later form a landmass composed of North America and Eurasia, and a southern part called *Gondwanaland*. In between the two was the Tethys Sea, considerably larger than the present-day Mediterranean, which played a major role in the creation of the Alps. In the Jurassic Period, 180 million years ago, when dinosaurs were the dominant land-animal life form, the nascent Atlantic began to split Pangaea apart, first at the level of the Equator then to the north and south. This rift separated what we now call North America from Europe, and South America from Africa.

Plate Collisions and Creation of Relief

In a dramatic about-face, the Afro-Asian crustal plate which up to that point had seemed to be moving away from ancient Europe began to rotate counterclockwise. As the ocean separating Europe from Africa retreated, the two plates came together in a monumental collision that at the end of the Mesozoic Era started to force up all the mountain ranges of southern Europe, from the Sierra Nevada in southern Spain to the Caucasus Mountains in southwest Russia. In between arose the Pyrenees in France and Spain; the Alps in south central Europe; the Apennines in Italy; the Pindus and the Peloponnese in Greece; and in the Maghreb, the Rif Mountains in northern Morocco, and the Atlas Mountains in northern Algeria.

In France, the collision between the two plates that marked the uplift of the Pyrenees in the Cretaceous (late Mesozoic) and in particular the creation of the Alps in the mid-Tertiary Period also caused a series of counter-shocks across the entire French landscape that was felt as far as the Gironde (in the southwest), the lower Loire Valley, the Coteaux du Layon region, and Normandy. The ancient bedrock braced by the Massif Armoricain, the Massif Central and the Vosges was forced up along the eastern edge of the Massif Central, and the Vosges, with faults opening up in numerous places. Some blocks were downthrown and others were upthrown, often in a north-south direction along ancient faults that scarred the European basement. Next came the forces of erosion. Huge quantities of rocks and stones were ripped from the newly formed peaks and carried for hundreds of miles, depositing in valleys and plains and along rivers. There is hardly a single area in southern Europe that was not directly affected to some degree. Millions of years later, the edges of faults, piles of scree and flat, stony stretches of land left behind by the Alpine upheaval would become ideal sites for viticulture.

Going Back through Time: The Geological Eras

Geologists have grouped periods of mountain building, ocean ebb and flow, and sedimentary deposition into three eras. By the beginning of the present era, the Earth's surface had been shaped and modeled to leave roughly the topography we see today.

Orogeny

Post orogeny

Terrestrial crust

Metamorphic rock

Granite

Volcanic rock

In the course of orogeny, magma seeped into the Earth's crust. Some of it welled up to the surface forming volcanic rock. Some of it remained deep within the Earth and crystallized to form pockets of granite. The surrounding rocks meanwhile metamorphosed in the intense heat.

Granite composé de feldspath, de mica et de quartz.

Paleozoic Era: Formation of Ancient Bedrock

The Paleozoic Era lasted for nearly 250 million years, or as long as the Mesozoic and Cenozoic eras combined. This is when the *ancient bedrock* was formed. The bedrock was then exposed to repeated cycles of mountain building that ended with the Hercynian Orogeny. Traces of the Hercynian Orogeny have survived in the relief and geology that we see today. Indeed, many ancient massifs are described as "Hercynian."

Sedimentary Rock

The Hercynian uplift was preceded by an accumulation of sediment. Remnants of these sediments are apparent in the Armoricain sandstone found in Brittany (northwestern France); the Brioverian (pre-Cambrian) schist found in the Nantes region (western France), famous for Muscadet wines; and the Silurian-Devonian schist and limestone found in the southern Massif Central, home to the northernmost vineyards of Saint-Chinian and Faugères (southeastern France).

Volcanic Rock

Throughout the periods of orogeny, pockets of molten magma built up in the Earth's crust. Some of this magma welled up to the surface to form volcanoes and volcanic rocks – in the southern Vosges Mountains of eastern France, for instance. The remaining magma crystallized at great depths within the Earth's crust to form *granite*. The vineyards of Saint-Joseph and Condrieu (Northern Rhône Valley) and Beaujolais-Villages (Burgundy) are rooted in granite.

Metamorphic Rock

Due to intense pressure and heat at great depth, the rocks surrounding the pockets of molten magma were metamorphosed. Examples of metamorphic rock include *micaschist* and *gneiss*, both found in the vineyards of the Côte Rôtie (Northern Rhône Valley).

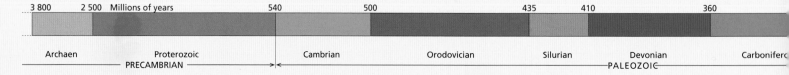

| 3 800 | 2 500 | Millions of years | 540 | 500 | 435 | 410 | 360 |

| Archaen | Proterozoic | Cambrian | Orodovician | Silurian | Devonian | Carbonifere |

← PRECAMBRIAN → ← PALEOZOIC →

The Muscadet and
Coteaux d'Ancenis region
(Loire Valley) was once an
ancient mountain range
attached to the Massif
Armoricain. It was largely
unaffected by the Alpine
upheaval but was
gradually reduced by
weathering and erosion
to a low plain known
as a peneplain.

Gneiss, a very common
form of metamorphic
rock composed of dark
zones (micas,
amphiboles) and light
zones (quartz, feldspar).

295		245			195		135		96		65	55		36	24		5	2
			Lower	Mid.	Upper	Lower	Mid.	Upper		Lower		Upper		Eocene		Miocene		
													Paleocene		Oligocene		Pliocene	
Permian				Triassic			Jurassic				Cretaceous							
						MESOZOIC								TERTIARY				
																QUATERNARY		

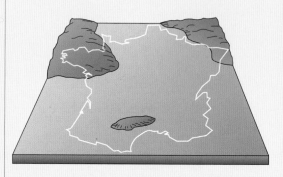

France in the Jurassic Period.

Sedimentary rock at the surface of the Earth can be altered into metamorphic rock under the intense pressure and heat of folding and compression associated with mountain building.

Peneplains

At the end of the Paleozoic Era the mighty Hercynian Mountains were worn down by weathering and erosion to a relatively flat surface called a *peneplain*. The ridges and valleys of the Massif Armoricain in Brittany are a prime example. As the forces of erosion gradually wore away the layers of the Earth's surface, pockets of granite and surrounding metamorphic rock were exposed. The ancient bedrock is now very much in evidence in Germany, Portugal, and South Africa but much less apparent in France, where it only outcrops in Brittany, the Vosges and Ardennes mountains (eastern and northeastern France, respectively), in the Massif Central and deep in the Alps and Pyrenees. It also underlies the entire Paris and Aquitaine basins (southwestern France) beneath the Mesozoic and Cenozoic layers. The whole of France in fact rests on ancient bedrock.

By the end of the Paleozoic Era, the surface of the Earth was considerably flatter and the Hercynian trend had been weathered and eroded. The famous Sandstone Vosges were formed from eroded relics of the Ardennes. During the middle Paleozoic), animal life emerged from the water and reptiles came to live on dry land.

Mesozoic Era: The Invading Sea

The Mesozoic Era was a period of calm following the Hercynian storm but leading up to the Alpine storm. It began with the Triassic Period when France was gradually invaded by sea. Sediment was deposited in coastal deltas and in warm saltwater lagoons in southeastern France, Aquitaine, and on the site of the future Paris Basin. The sea was deeper in the east, around the Alps and northern Italy.

Marine Deposits

In the Jurassic Period, the sea submerged most of the land surface in France. The only remaining areas of dry land were present-day Brittany, the Montagne Noire (southeastern France), and the Ardennes. Substantial limestone deposits were laid down on the seabed by colonies of coral in the shallow zones to the east of the Paris basin and in Burgundy. Limestone was also deposited through natural sedimentation.

Gradually the layers of marine deposit in the Paris Basin grew so thick and heavy that the center of the basin started to subside under the extreme weight of sedimentation. This is the phenomenon known as *subsidence*. Depending on environmental conditions (climate, sea depth, proximity to the shoreline), limestone alternated with predominantly hard sandstone, marls, and sands or softer clays.

The arrangement of each formation had a decisive influence on regional topography. We find evidence of the mid-Jurassic in the eastern Paris Basin where deposits of very hard limestone left by coral reefs alternate with softer marls. Other mid-Jurassic formations include the Kimmeridgian that underpins the vineyards of Chablis, Sancerre, Menetou-Salon (Loire Valley), and Champagne (Aube); and the Bajocian, Bathonian, and Oxfordian limestone on the Côte d'Or. In the same period, the Aquitaine Basin split into two. To the east were shallow waters and coral reefs that left huge banks of limestone, such as we find in the Quercy Plateau on the southeast flank of the Massif Central; to the west and out at sea were more marly limestone deposits that would later form the

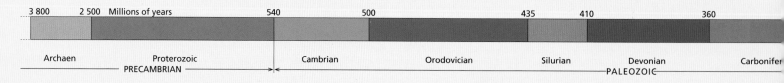

3 800	2 500	Millions of years	540	500		435	410		360
Archaen	Proterozoic		Cambrian	Orodovician		Silurian	Devonian		Carbonifer
	——— PRECAMBRIAN ———						——PALEOZOIC——		

The Aÿ vineyard in Champagne is rooted in Campanian chalk.

subsoil of the Cognac region.

This is the period when the nascent Atlantic began to split Europe and Africa from America. As the Pacific plate began to descend beneath the American continent, it created the Andes in South America and volcanoes that would form the basis for the Sierra Nevada in North America.

The African Plate Changes Direction

In the Lower Cretaceous Period (135 million years ago) the sea retreated from practically the whole of France. The end of this period coincided with a major geological upheaval that began a few hundred miles to the southeast of France and would change the face of southern Europe forever. All at once, the ocean separating southern Europe from Africa stopped expanding. The African plate switched direction and headed toward Europe.

At the end of the Lower Cretaceous, the sea returned to the Paris Basin and the Rhône Valley, depositing a very hard, dense reef limestone called Urgonian limestone, found today around the vineyards of the southern Rhône Valley. Further south, the northward migration of the African plate sent the first shock waves ricocheting through Provence.

Calm was restored in the Upper Cretaceous, and things remained quiet for the next 30 million years or so. The Cenomanian and Turonian were times of widespread geological transgression (a relative rise in sea level resulting in deposition of marine strata over terrestrial strata). The sea invaded the Paris Basin, where it deposited the thick layers of chalk after which the Cretaceous is named. The Champagne vineyards in the Marne Valley and the vineyards in the Touraine and Saumur are rooted in soils derived from Cretaceous chalk. The sea also returned to the Aquitaine Basin, which extended along a narrow gulf as far as Provence. But the approaching African plate, although still at a safe distance, now made itself felt once again. The Pyrenees Mountains grew. There was crustal deformation in the Iberian Peninsula. By the time the Mesozoic gave way to the Cenozoic Era (65 million years ago), the dinosaurs were on the verge of extinction and the sea had retreated from most of France. In California, the Laramian orogeny triggered the uplift of the Sierra Nevada Range and the creation of the Coast Ranges, which would later become home to the vineyards of the Napa and Sonoma Valleys. In South America, the Andean uplift continued, becoming more intense in the Tertiary Period.

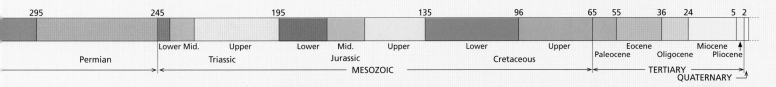

| 295 | | 245 | | 195 | | 135 | | 96 | 65 | 55 | 36 | 24 | 5 | 2 |

		Lower	Mid.	Upper	Lower	Mid.	Upper	Lower	Upper		Eocene		Miocene	
Permian			Triassic			Jurassic		Cretaceous		Paleocene		Oligocene		Pliocene
						MESOZOIC					TERTIARY			
												QUATERNARY		

Argentina: stones washed down by rivers rising in the Andes are scattered along the way and then deposited at the foot of mountains, providing an excellent medium for viticulture.

Tertiary Period:
A Time of Uplift and Subsidence

Things quieted down in the earliest epoch of the Tertiary, the Paleocene. The sea in the Paris Basin continued its steady retreat from the center. In the Eocene, the migrating African plate caused renewed compression and tremors. The Corsica-Sardinia landform bulged upward, thrusting its sedimentary layer northward, particularly toward Aix-en-Provence (southern France). At the time, Corsica and Sardinia were not where we find them today but alongside the Pyrenees and the Maures, which together formed a single mountain range.

Formation of the Rhine Graben

Like a huge bulldozer, the African plate shoveled large quantities of material toward northern Italy. The Alpine Sea moved westward and shrank. Even the Paris Basin was shaken up in the collision. Then in the Oligocene,

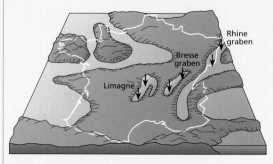

France in the Oligocene.

things quieted down again. The intense pressures diminished and the earth entered what geologists call a *relaxation phase*. Like an object slipping out of a vise that has lost its grip, Alsace collapsed between the Vosges and the Black Forest—which were not yet mountainous—creating a strait between the north and the south seas. In the same way, La Bresse and the Limagne (Clermont-Ferrand and Roanne-Saint-Etienne basins) also collapsed, as did the Swiss Plateau. In the

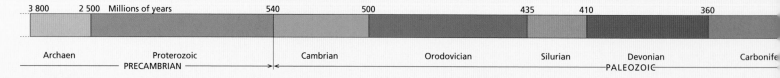

3 800	2 500	Millions of years	540	500	435	410	360

Archaen	Proterozoic	Cambrian	Orodovician	Silurian	Devonian	Carbonife
	PRECAMBRIAN				PALEOZOIC	

Paris Basin the sea launched a final invasion that left the sands of Fontainebleau, renowned for their high-purity silica.

Valleys Filled by Erosion

In the Aquitaine Basin meanwhile, the sea retreated almost completely, remaining only around the Landes. The young Pyrenees were subjected to intense erosion, shedding materials that accumulated on the northern and southern flanks, first in the sub-Pyrenean trough along the edge of the mountains, then throughout the southern part of the basin up to the Bordeaux-Castres level. These sediments formed the molasse such as we find in the vineyards of Armagnac. In the early Miocene epoch, the sea deposited sand and marls in the future Rhône Valley and as far as Bresse, filling the graben. This process was repeated in Alsace and the Limagne.

Calm before the Storm

The African plate, shoving the accumulated debris of 60 million years ahead of it, finally reached the Alps. The granite bedrock that until then had lain beneath the Alpine sea smashed into the Massif Central and was thrust upward to form the Mont Blanc and Pelvoux mountain peaks. The layers of sediment deposited on the bedrock in the Jurassic and the Cretaceous slipped down the western flank of the new range, creating the limestone ranges of Aravis, Bauges, La Chartreuse, and Vercors. Farther north, the same layers folded to form the Jura where the ridge overlooking La Bresse would later become home to the Jura vineyards.

The ancient Hercynian bedrock did not survive the impact. At the end of the Miocene, the entire eastern edge of the Massif Central from the Cevennes to the Morvan, together with the eastern edge of the Vosges, rose up like a giant swelling. Magma spewed out along the north-south faults breaking up the ancient mountain range and gradually forming the chain of puys (peaks

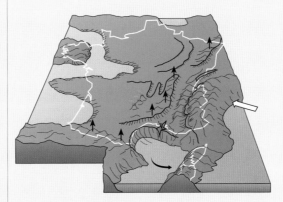

France in the late Miocene.

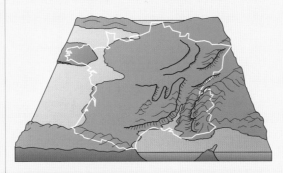

France in the Pliocene.

of volcanic origin) that are still developing today.

The newly-formed Alpine peaks were severely eroded, shedding vast amounts of material that accumulated along the western edge of the mountains, forcing the Rhône River back into a narrow passage at the foot of the Massif Central. In the Mediterranean, the Corsica-Sardinia landform rotated counterclockwise around the northern tip of Corsica. The Gulf of Lion opened up and collapsed. The Mediterranean Sea, having lost touch with the Atlantic, dried up and fell to more than 6,000 feet below present sea level. The watercourses flowing down the Alps and the Massif Central carved deep ravines in the limestone plateaus of southeastern France, sculpting the gorges of the Ardèche and Verdon and the narrow creeks (*calanques*) in the cliffs of Marseilles. In Italy, the Po plain collapsed. During the Pliocene epoch, the Mediterranean Sea returned to its previous level.

Main stages of relief formation in France. Jurassic (p. 16): the entire territory is invaded by sea, which leaves thick sediments. Oligocene (p. 18): the Paris Basin remains underwater; the valleys of the Allier, the Loire, the Rhine and the Saône collapse. Miocene: the Alpine Storm reaches its height. The Corsica-Sardinia landform tilts toward the southeast and the Mediterranean dries up. Pliocene: the sea returns to invade the Rhône Valley, and the Atlantic Ocean launches a final invasion in the Lower Loire Valley.

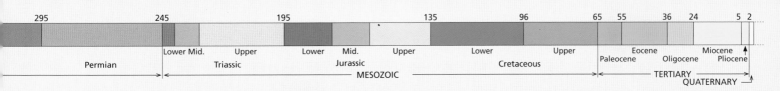

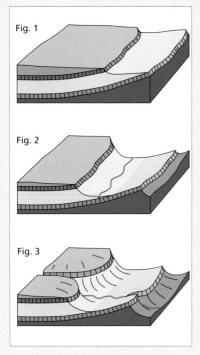

Cuesta formation in the eastern Paris Basin.

Fig. 1: **Vosges uplift leads to slightly sloping strata.**

Fig. 2: **A river cuts through the softer layers, eroding the face of the more resistant strata (vertical hatching).**

Fig. 3: **The uplift becomes more pronounced in the east. The river now flows toward the center of the basin and cuts a notch in the resistant rock (e.g. Marne River in Epernay and Aube River in Bar-sur-Aube, Champagne region).**

The four main glacial periods of the Pleistocene epoch (Qauternary Period)

Quaternary Period:
Introducing Landscapes

The landscape finally took shape in the Quaternary, as glacial processes and erosion chiseled the mountains and sedimentary layers and dumped deposits of waste rock in the valley bottoms. Erosion has been a feature of geological evolution since time began, thanks to the leveling effect of gravity, which wears away high areas and so provides material that accumulates in low areas. The Pyrenees and the Alps have been at the mercy of this process since they were first formed. In France, however, erosion and deposition have been far more significant for the past few million years because virtually the entire French landform was uplifted at the end of the Tertiary. Witness the uplift of the Pyrenees, Alps, Corsica, Jura Mountains, and Massif Central from the Morvan to the Montagne Noire. Even the edges of the Paris and Aquitaine basins were raised. Subsequent erosion in the Quaternary added the final touches to each hill, talus slope, and plain on which today's great vineyards are planted. Some details took longer than others: the Muscadet, Alsace and Côte Rotie terroirs were already in place at the start of Quaternary, but the Médoc Graves and southern Rhône Valley terroirs had yet to emerge.

Terroirs Sculpted by Water

Water in all its forms is the chief agent of erosion and leaves its mark on the landscape in every way possible. Rainwater breaks up rocks by dissolving some of the minerals they contain. Ice is just as effective: water seeps into rock fractures and then freezes at night, causing the rock to disintegrate. Torrents and rivers wear away the lands through which they flow, and this form of erosion is the most interesting in the study of terroirs.

Softer formations such as marls, marly limestone and sands are the first to be washed away, exposing the

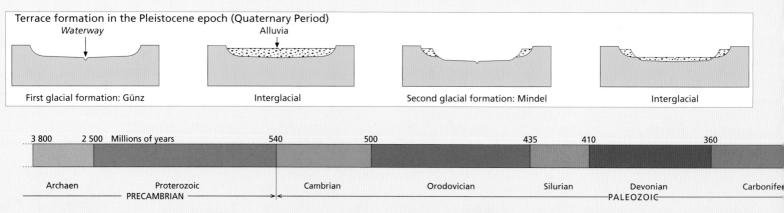

Terrace formation in the Pleistocene epoch (Quaternary Period)

Waterway	Alluvia		
First glacial formation: Günz	Interglacial	Second glacial formation: Mindel	Interglacial

3 800	2 500	Millions of years	540	500	435	410	360

Archaen	Proterozoic	Cambrian	Orodovician	Silurian	Devonian	Carbonifer
PRECAMBRIAN				PALEOZOIC		

hard limestone layers beneath. Eventually even these more resistant rocks may be affected. One of the best examples of such erosion is the famous series of outward-facing scarps (cuestas) that rim the Paris Basin in northeastern France. Here, layers of resistant rock cover softer types of rock in which the vines take root. In the Côtes d'Auvergne, flows of hard, water-resistant basalt protected the subjacent terrain underlying the vineyards of Châteaugay and Corent. Farther downstream, the Loire made its bed in the Cretaceous rocks of the southern Paris Basin; the vineyards of Bourgueil and Chinon are planted on slopes etched out by the Loire then covered by the tuffeau of the Turonian.

Vineyards are frequently planted along major rivers. The reason for this, say geographers and historians, is that waterways provided a means of transport and communication, vital to the development and ultimate success of speculative plantings such as vines. Geologists highlight the quality of terroirs created by fluvial erosion in the course of the Quaternary glaciations. The uplifting of mountain ranges produced today's major rivers. The Garonne and Rhône and their tributaries, for instance, originated in the nascent Pyrenees and Alps, respectively. They then scoured the slopes, carrying masses of material that was rolled, worn away, and ground down before eventually being dumped where the slope leveled out--leaving vast, shelf-like surfaces covered with fill. These stony terraces are usually particularly suitable for vine growing. Essentially, however, the quality of a terroir is de-termined by the rock fragments that are found there,

which is another way of saying that terroirs of quality are located in regions eroded by rivers.

Terraces Born of Glaciation

The Quaternary marked a series of glacial periods alternating with warmer periods (interglacials) that accounts for the arrangement of the alluvial terraces found in France. In one glacial period, water turned to ice, sea level fell and valleys were scooped out of the Earth's surface. The following interglacial brought a thaw, rising sea levels, shallower sloping channels and alluvial deposition. These alluvia were scooped out in the next glacial period (as rivers dug deeper channels) and replaced with a fresh layer in the next interglacial, and so on. This continued downcutting left successive levels of terraces that correspond to the four main glacial stages of the Pleistocene Ice Age: the Günz, Mindel, Riss, and Würm. Terraces are a frequent sight along rivers, and in France they became home to some of the country's most celebrated vineyards (Graves, Médoc communal appellations, Côtes-du-Rhône, Châteauneuf-du-Pape, etc). In the course of the Quaternary, glaciers became far more widespread than they are today and reached as far as the Alpine valleys. Glaciers, like rivers, scraped and rasped the sides of mountains, pushing material along and dumping it ahead of or alongside the glacier in moraines. When the glaciers retreated, these frontal and lateral moraines became home to vineyards in Haute-Savoie and in Switzerland.

The big round stones (galets roulés) characteristic of Villafranchian terraces are famously associated with Châteauneuf-du-Pape.

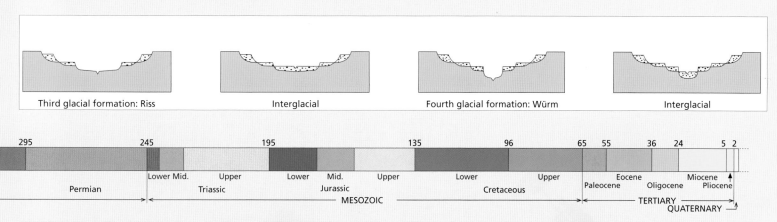

| Third glacial formation: Riss | Interglacial | Fourth glacial formation: Würm | Interglacial |

| 295 | 245 | 195 | 135 | 96 | 65 | 55 | 36 | 24 | 5 | 2 |

| Permian | Lower Mid. | Upper | Lower | Mid. | Upper | Lower | Upper | Paleocene Eocene | Oligocene | Miocene Pliocene |
| | Triassic | | | Jurassic | | Cretaceous | | | | |

MESOZOIC — TERTIARY — QUATERNARY

Vineyards on the Edges of Faults

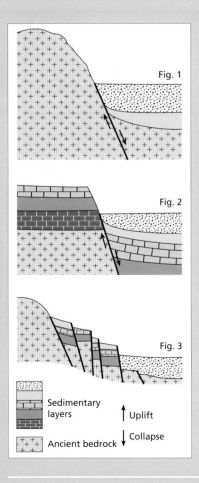

Fig. 1

Fig. 2

Fig. 3

Sedimentary
layers

↑ Uplift

↓ Collapse

Ancient bedrock

Imagine a vineyard and you see rows of vines planted on sunny slopes. In temperate regions, vineyards are predominantly located on east-, south- or west-facing slopes where the vines receive maximum sunshine but are less likely to suffer frost damage than on plains or valley bottoms.

These slopes were often created by faults associated with the formation of the Alpine ranges in the Tertiary Period. They were produced in a two-stage process. In the mid-Tertiary (Oligocene), a relaxation phase led to the collapse of Alsace, the Saône and Bresse Valley, the Lower Dauphiné (a former province of southeastern France), the Limagne (valley of the Loire and its tributary the Allier) and the Swiss plateau. These downthrown blocks created topographic troughs, or grabens. Then in the late Tertiary (Miocene) the ancient bedrock was forced up by the Alpine upheaval, leaving lasting marks on the Massif Central and the Vosges. These two movements happened in reverse along ancient faults that scarred the Hercynian bedrock. Subsequent tectonic movement reactivated these faults, shearing through whatever sedimentary strata may have been overlaid.

The amount of slippage or fault throw (vertical displacement of rock strata at a fault), the number of faults, the presence or absence of sedimentary strata, and the type of sedimentary strata together determine very different geological locations and, therefore, very different terroirs.

The most geologically simple locations are found in areas of pronounced uplift, where the ancient bedrock is exposed along the top of the slope and any overlying sedimentary strata have generally been worn away by erosion. This is the case in the Côte-Rotie, Condrieu, and Saint-Joseph appellations in the Rhône Valley, and in the Côte-Roannaise appellation in the Loire Valley.

In places where the uplift was less pronounced, the sedimentary strata may have survived; these are the strata visible on the hillside, along the fault. The Côte de Nuits and Côtes de Beaune (Burgundy), with their relatively homogeneous formations, are good examples of such locations. The most complex locations are those criss-crossed with a whole network of more or less parallel, adjacent faults that create a patchwork of very different terroirs, as in Alsace, for instance.

Alsace

The Alsace vineyards extend for more than 60 miles along a narrow ribbon of land running through the sub-Vosgian Hills and the immediately adjacent lands of the Rhine Plain. The vines here enjoy a privileged location, shielded by the Vosges Mountains that deflect cold winds and rain from the ocean. The climate is semi-continental. Wine connoisseurs tend to distinguish Alsace wines by grape variety, but one of the region's most outstanding features is the sheer diversity of terroirs arising from the many fault zones. This variety results in numerous microclimates that bring out the characteristics in different grape varieties. Since 1975, when the first Alsace Grands Crus were classified, a total of 50 vineyards have now been classified.

The slopes supporting the terroirs of Alsace developed in two stages: collapse of the Rhine Plain and uplift of the Vosges Mountains. The same phenomenon occurred along the edge of the Black Forest in Germany. The Vosges uplift was much more pronounced on the Alsace side, in the east, than on the Lorraine side, in the west.

Around Two Faults

The vine-planted slopes in the east that overlook the Plain of Alsace did not develop from a single fault but from a multitude of parallel cracks running almost north-south. They are flanked by two major faults: the Vosge Fault in the west, along the ancient bedrock; and the Rhine Fault in the east, alongside the plain.

• The massif in the west consists of essentially granitic primary terrain that remains partially covered with Triassic sandstone. All trace of Mesozoic and Tertiary layers has, however, long since disappeared.

• The sub-Vosgian Hills, separated from the mountain by the Vosges Fault, were produced by a series of stepped faults that cut out slopes running north-south and east-west. Imagine a somewhat dilapidated Roman amphitheater, with tiered and staggered seating. Step faulting introduces successively younger strata: the sedimentary layer survives on the section of block that has descended into the fault, whereas the strata that outcrop at the surface differ depending on each block's degree of collapse. This patchwork of strata, ranging from Oligocene to Triassic in age, produced a complex variety of soils. For the most part, however, the soils are derived from clays and limestone. Moreover, due to slope erosion, the surface elements of a block are often mixed with those of the block directly underneath.

• The *alluvial fan* to the east of the Rhine Fault contains Tertiary sediments overlain with alluvia stripped from the Swiss Alps by the Rhine River, and by alluvial cones formed by rivers as they debouch from the Vosges (Giessen Valley in Sélestat; the alluvial fans of the Fecht in Wintzenheim, and of the Weiss in Kientzheim, west of Colmar).

The Three Terroirs of the Alsatian Slopes

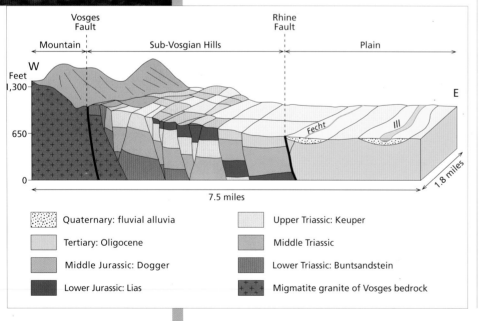

Vines are in their element in the sub-Vosgian Hills and it is here that the majority of the vineyards are located. They are less common in the mountains or in the alluvial fans of rivers flowing from the Vosges. However, this overall pattern of three strips separated by two faults does not apply throughout the region. In fact, around Sélestat, Colmar, and to the south of Guebwiller, the Vosges and Rhine faults merge, bringing the ancient Vosgian bedrock into contact with the plain. The slopes so formed are some of the steepest wine-producing slopes in France.

Soils

The sub-Vosgian Hills are home to a vivid patchwork of surface formations created by fault blocks at different heights. Some of these formations proved more resistant to erosion than others and form ridges, often crowned by picturesque villages. A good example is Zellenberg, east of Riquewihr, where conditions on the south and southeast sides are particularly suited to viticulture, as shown by the number of Grands Crus. The vineyards are located at heights of 650-1,150 feet, rising terrace by terrace. The terroirs of the sub-Vosgian Hills boast a variety of soils, containing marl and limestone.

• Marly/argillitic: heavy, fertile soils derived from Keuper marly formations and Lower and Middle Jurassic strata; located mainly in the Saverne and Ribeauvillé fracture zones. They are often overlain with Quaternary scree that improves soil aeration. Notable terroirs located on such soils are the Grands Crus of Riquewihr, Schoenenbourg, and especially Sporen, one of the rare Grands Crus planted on a shallow slope.

• Calcareous, sandy: light, well-aerated soils, derived exclusively from Triassic sandstones and sometimes Tertiary strata. This type of soil is confined to relatively small areas, such as the Zinnkoepflé Grand Cru, near Soultzmatt and Westhalten, that faces full south from the top of a majestic slope.

• *Limestone*: dry, very stony soils produced from Mid-Triassic and Lower Liassic limestone terrain and Great Bajocian Oolite (Middle and Lower Jurassic). These soils provide a favorable medium for viticulture around Saverne, on the edges of the

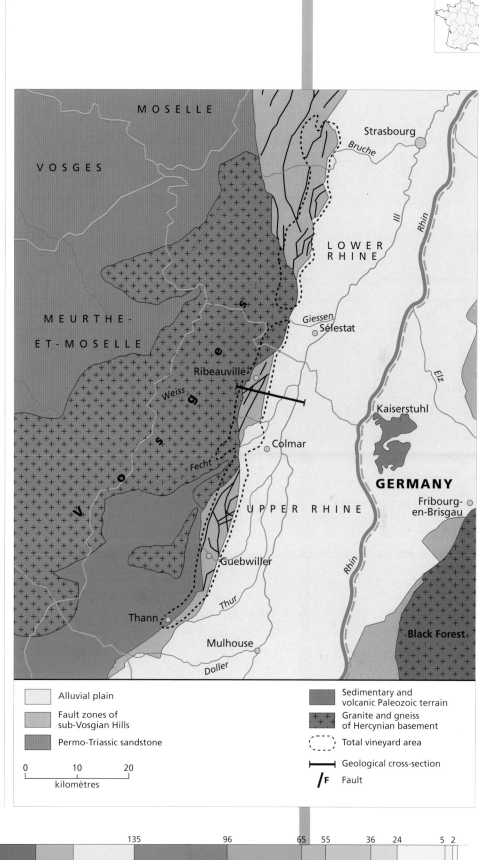

Alluvial plain

Fault zones of sub-Vosgian Hills

Permo-Triassic sandstone

Sedimentary and volcanic Paleozoic terrain

Granite and gneiss of Hercynian basement

Total vineyard area

Geological cross-section

/F Fault

0   10   20
kilomètres

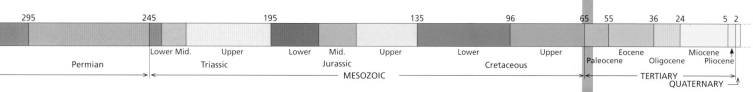

295		245		195			135	96	65	55	36	24	5 2

Permian — Triassic (Lower, Mid., Upper) — Jurassic (Lower, Mid., Upper) — Cretaceous (Lower, Upper) — MESOZOIC — Tertiary (Paleocene, Eocene, Oligocene, Miocene, Pliocene) — QUATERNARY

The Schoenenbourg
Alsace Grand Cru
(Riquewihr commune).

Facing page:
The Schlossberg
Alsace Grand Cru
(Kientzheim commune).

• *Volcanic sedimentary*: rather shallow but fairly rich soils found along the southern border of the Vosges on steep slopes that drop down to the plain. One of the best examples of this type of terroir is the Rangen Grand Cru overlooking the town of Thann.

• *Granite and gneiss*: *arenitic* soils from weathered rocks, found at varying depths depending on the slope. They occur along the edges of the Vosges, to the north and south of Sélestat (Wintzenberg and Frankstein Grand Crus) and on the upper western border of the Ribeauvillé fracture zone where they support the Grand Crus of Schlossberg at Kientzheim and Brand at Turckheim, northwest of Colmar.

• *Alluvial*: sandy pebbly soils with localized patches of clay found in the alluvial fans of rivers as they exit the Vosges, especially the Weiss and the Fecht. The quality of these terroirs, here as elsewhere, depends on the minerals leached from the soil.

Grape Varieties

Prior to the phylloxera epidemic, Alsace was home to a wide variety of cultivars, but today only three groups of grape varieties are authorized for production as Alsace appellation wines: traditional cultivars (Riesling and Chasselas); Burgundian varieties (Pinot Noir, Pinot Blanc, and Pinot Gris); and Swiss and Austrian vine stock. This last category includes the Traminer, Klevener de Heiligenstein, Gewürztraminer, Sylvaner, first planted in Alsace in the nineteenth century, and Muscat Ottonel, established in the twentieth century. Of these varieties, only the Gewürztraminer, Pinot Gris, Riesling, and Muscat qualify for Grand Cru production. They may also be made into the sweet Alsace appellation wines, Vendanges Tardives and Sélection de Grains Nobles, using late-harvested grapes that are almost bursting with sugar.

Growers in Alsace have been vinifying grape varieties separately for more than one hundred years, so we can see here how different grape varieties respond to different types of terroir – noting of course that it is not always possible to match the cultivar to the soil. The Muscat and Pinot Noir varieties, for

Rouffach region and in some of the Ribeauvillé terroirs.

• *Marly limestone*: characteristically stony soils at varying depths and more or less calcareous, composed of a mixture of marls plus large quantities of pebbles of diverse origin from eroded slopes. These terroirs are naturally favorable for wine growing and occur predominantly along the Oligocene formations on the eastern side, laid down by the weathering of upstream strata.

• *Mountainous*: these soils support vineyards at the junction with the plain near Seléstat, in the upper belt of the sub-Vosgian Hills, and on the sides of valleys. An example is the celebrated Schlosserberg Grand Cru that overlooks the River Weiss, below Kaysersberg.

• *Sandstone*: light, sandy, permeable, rather lean soils, principally located south of Saverne in the Andlau region and in a small area of Guebwiller.

• *Schistous*: clayey soils confined to the terroirs south and north of Andlau (Kastelberg Grand Cru).

3 800	2 500	Millions of years		540		500		435	410		360	
Archaen		Proterozoic			Cambrian		Orodovician		Silurian		Devonian	Carbonifer
		PRECAMBRIAN								PALEOZOIC		

instance, have been shown to thrive in limestone terrain. The Gewürztraminer prefers marl and limestone soils, which give the wines body, fruitiness, and good aging potential. Good examples of this type of terroir are the Mambourg and Furstentum Grands Crus, north of Sigolsheim. Gewürztraminer planted in granitic soils yields fine, elegant wines with floral and exotic notes. They have less body than wines produced from marly limestone terrain but grow more expressive with time. Gewürztraminer wines originating from alluvial soils are earlier maturing, lightly fruity, and should be drunk "young" (relatively soon after the wine is produced).

The Riesling, king of the Alsace varieties, is at home almost anywhere but behaves quite differently depending on the soil. Riesling wines originating from granitic soils, such as those typical of the Brand Grand Cru near Turckheim, are fruity, with exceptional finesse, and suitable for early drinking. In contrast, Riesling wines produced from grapes planted in limestone or marl and limestone are more spirited and less fruity but have added power. Riesling vines cultivated in alluvial soils, like

Gewürztraminer vines, yield early maturing, light, and fruity wines. Those planted in the schistous terroirs of Andlau have a distinctive musky aroma.

The Riesling

Whether you regard the Riesling as a great Alsatian or a great German cultivar rather depends on what side of the Rhine River you come from. To wine connoisseurs, it is simply one of the world's greatest white grapes. The vineyards of Alsace, virtually the only part of France to cultivate the Riesling, lie at the southwest tip of a much larger area of Riesling acreage centered in Germany. Plantings there account for 21,000 hectares of the total area under vine, compared with just 3,000 in Alsace. The Riesling is nevertheless the premier grape of Alsace, more widely planted than the Gewürztraminer and adaptable to almost any type of soil.

The Kitterlé Alsace Grand Cru: a sandstone terroir, mainly planted with Riesling vines.

Most of the Alsace Grand Crus can be related to a single geological terrain. In order to see which terroirs the Riesling prefers, we have only to look at the string of famous vineyards dotted along the slopes of the Bas-Rhin.

A Preference for Granitic Sands

Beginning in the north, we find a predominance of Triassic sandstone formations between Marlenheim and Molshei planted with Riesling and Gewürztraminer vines. The Riesling shows a particular preference for the very stony Liassic marly limestone terrain of the Altenberg de Wolxheim Grand Cru. Slightly farther south, more plantings of Riesling occur in the stony Oligocene conglomerates of the Zotzenberg Grand Cru near Mittelbergheim, and especially in the terroirs from Andlau to Damback-la-Ville, where the Paleozoic basement meets the Plain of Alsace. The vineyards here are mainly devoted to the Riesling that thrives in Paleozoic granitic schists and sands (Kastelberg and Winzenberg), Triassic sandstone (Wiebelsberg), and gneiss in Praelatenberg, the final Grand Cru in the Bas-Rhin.

The Riesling declines in the northern part of the Haut-Rhin but then returns with a flourish in the granitic sands that support the Grands Crus of Schlossberg (Kientzheim), Katzenthal, Niedermorschwihr, and Brand de Turckheim. It fades out again south of Colmar, and then re-appears in the Pfingstberg d'Orschwihr Grand Cru, around Guebwiller, where it thrives in Triassic sandstone. At the southern end of the Alsace Wine Route, the Riesling acquires exceptional finesse in the unique volcanic terroir of the Rangen Grand Cru.

Riesling Characteristics

The Reisling evidently favors the terroirs situated on the Vosges bedrock and overlying Triassic strata, especially granitic, sandy soils. The wines originating in such soils are fine and elegant, with a capacity to open up in their youth. Their rich bouquet of floral and fruity aromas (white fruits) reveals mineral nuances that can become dominant after just a few years' cellaring. Riesling wines originating in schists (Kastelberg Grand Cru)

and volcanic terrain (Rangen) are particularly distinctive. The stoniest marly limestone soils yield subtly floral Riesling wines with a hint of citrus. They tend to be less exuberant, slower to open but more powerful in the long run. Riesling wines originating in the stony soils of alluvial fans are light and quaffable but shorter-lived.

Steeply Sloping German Vineyards

A tour of Riesling vineyards would not be complete without a look at the terroirs on the German side of the Rhine. In Germany, the Riesling is the second most popular grape variety after the Müller-Thurgau, a less demanding Riesling-Sylvaner hybrid that accounts for 24,000 hectares of the area under vine, compared with 21,000 for the Riesling. The finest German wines, however, all originate in the Riesling that we find mainly planted along the valley of the Rhine and its tributaries, the Mosel, Nahe, and Neckar, sometimes in steeply sloping vineyards. Slower to ripen in Germany, the Riesling requires sunny, sheltered conditions. Streams and rivers reinforce the role of the soil by reflecting sunshine, while also providing useful moisture.

A wide variety of German wines are based on the Riesling, from dry to sweet wines (*Auslese, Beerenauslese, and Trockenbeerenauslese*) and the magnificent ice wines (*Eiswein*). Their characteristics are quite unlike those of French Riesling wines, for reasons mainly to do with the weather. German Rieslings are lower in alcohol, markedly more acidic, and superbly light and elegant at their peak.

The Sommerberg Alsace Grand Cru, behind the village of Niedermorschwihr: a sloping, granitic sandy terroir that extends into the neigboring village of Katzenthal.

The Gewürztraminer

No other grape variety is more characteristic of Alsace than the Gewürztraminer. More than a third of the world's plantings of the Gewürztraminer are concentrated in this single French region. Unlike the Riesling, which is mainly planted in southwest Germany, the Gewürztraminer accounts for less than one percent (830 hectares) of German acreage. The rate of cultivation is about the same in Austria and Australia (from 600-800 hectares) but significantly higher in the United States (1,000 hectares).

In France, setting aside a few areas in the Moselle (northeast France) that are actually extensions of vineyards in Luxembourg, the Gewürztraminer is grown exclusively in Alsace, where it accounts for 2,500 hectares of plantings, second only to the Riesling.

A Preference for Marls and Limestone

As you follow the Alsace Wine Route from north to south, the Gewürztraminer seems to be playing hide and seek with its rival, Riesling. One or the other tends to dominate in many of the Grands Crus, although in some *lieux-dits* (named vineyards) the two varieties co-exist in perfect harmony. Generally, Gewürztraminer grows best on Oligocene marls and limestone but also thrives in other terroirs. In the Bas-Rhin, from Marlenheim to Molsheim, the Gewürztraminer and Riesling share the limelight, growing on Muschelkalk marls and limestone and Keuper dolomitic marls. Slightly farther south, at the point where the ancient Paleozoic bedrock overlain by Triassic sandstone meets the Plain of Alsace, the Riesling becomes the predominant variety from Heiligenstein onward, growing equally well on granitic and gneissic arenitic terrain and on Triassic schists and sandstone. The Gewürztraminer makes a comeback in the Triassic sandy soils of the Gloeckelberg Grand Cru and the distinctly stony Jurassic marls and limestones of the Altenberg de Bergheim. However, it really comes into its own in the Zellenberg and Riquewihr terroirs, at the point where the sub-Vosgian hills are at their widest. In the south- and southeast-facing Froehn Grand Cru, for example, we find the Gewürztraminer flourishing in schistous marly vineyards at the foot of the hilltop village of Zellenberg. It gives equally good results in the Liassic marls of the gently sloping Sporen vineyard (Riquewihr).

Further south, the Gewürztraminer is fully at home in the Oligocene marly calcareous soils of the Sonnenglanz de Beblenheim vineyard, the Marckrain de Bennwihr vineyard, and the steeply sloping Mambourg Grand Cru, on the south-facing slope of Mont Sigolsheim. This particular vineyard was the focus of research by the French Institut National de la

The Gloeckelberg Alsace Grand Cru, on the border between the Bas-Rhin and the Haut-Rhin: the Gewürztraminer and Pinot Noir share acidic sandy soils derived from Vosgian sandstone.

The Sporen vineyard south of Riquewihr: the Gewürztraminer shows a particular preference for Liassic marls, even on gentle slopes such as the ones shown here.

Recherche Agronomique, examining the influence of the soil on grape behavior. The Gewürztraminer was shown to prefer soils that warm up quickly in the spring and provide a constant, moderate water supply throughout the year. This highly aromatic grape also dominates the vineyards adjacent to the village of Ammerschwihr, including the Kaefferkopf Grand Cru, where producers may occasionally mix grape varieties in the traditional manner.

Further south, the Gewürztraminer retreats from the vineyards of Katzenthal and Niedermorschwihr, where the Paleozoic basement once again contacts the plain. The grape then reappears in force south of Colmar on a fresh supply of Oligocene marls and limestones and tends to remain the dominant variety as far as Guebwiller, especially in the vineyards of Goldert, Steinert, and Zinnkoepflé. The Riesling then takes over at the southern tip of the Alsace slopes, from Guebwiller to Thann. Broadly, the Gewürztraminer prefers deep, marly soils to shallow, stony soils.

Riesling Characteristics

The Gewürztraminer's foremost feature is its rich, spicy bouquet of exotic fruits, often with musk-like overtones of geranium and rose that develop into notes of apricot and dried fruits in the dessert wines. Gewürztraminer wines that come from vineyards in alluvial fans are fruity, exceptionally light, and intended for early drinking. Those from granitic or sandy soils are robust but elegant and take longer to come into their own.

The Gewürztraminer performs best in the marl and limestone terroirs of the sub-Vosgian Hills, where it develops a depth of quality unmatched elsewhere, with the bouquet, body, and aging potential that are the mark of great Gewürztraminer wines.

Germany

Germany lies at the northernmost limit of vine cultivation, where the continental climate is rugged and marked by harsh winters. Conditions are even more challenging than in Switzerland, forcing producers to seek out the sunniest, most favorably exposed sites, sheltered from cold northern influences.

The majority of German vineyards are located along the Rhine Graben in southwestern Germany, flanked on the right bank by the rivers Neckar (Wurtemberg) and Main (Franconia) and on the left by the rivers Mosel and Ahr. This represents the principal axis of German viticulture and accounts for 60 per cent of total planting acreage. There are also a few minor vineyards in Saxony and in the mid-valleys of the rivers Unstrut and Elbe and its tributary the Saale.

A Geological History Similar to Alsace

The Rhine Graben and Alsace have a similar geological history and, as can be seen from the topography, the Alpine upheaval had repercussions for the entire German wine-growing region. The region remained submerged until the end of the Mesozoic Era and accumulated rich marine deposits. Then came a dramatic facelift in the Tertiary Period, which created topographical features favorable to viticulture.

At the end of the Eocene, and even more so in the Oligocene, what is now the Rhine Graben began to sink at its southern end, together with Bresse, the Bas-Dauphiné, and the Limagne in France. The sea invaded the land and left marls rich in potash salts that are still mined today in Alsace. After a brief period of calm, the Alpine upheaval intensified and the process of collapse resumed in the Miocene, especially in the north toward Frankfurt.

The graben was by now a distinctive feature, becoming more pronounced as a result of the reverse movement in the Pliocene and the Quaternary that uplifted the two margins of the graben: the Haardt and the Bergland, to the west of the Rhine; the Black Forest, Kraichgau, and Odenwald, to the east. The

Above:
Cochem in the
Mosel Valley.
Right:
The Scharzhof vineyard
in Wiltzingen,
Sarre Valley.
Top right: Schloss
Johannisberg in
the Rheingau.

3 800	2 500 Millions of years		540	500		435	410		360
Archaen	Proterozoic		Cambrian	Orodovician		Silurian	Devonian		Carbonife
	← PRECAMBRIAN →						← PALEOZOIC →		

rising Paleozoic bedrock ripped through the overlying Mesozoic strata, baring vast stretches of Paleozoic and Triassic terrain. The Rhine River, between Mayence and Cologne, and the Mosel River gradually cut into the deep slate and shale to form the valleys so favorable to viticulture.

There was also a certain amount of volcanic activity, mainly in the Miocene. Basalt magma bled out of the cracks in the graben, forming the Vogelsberg to the north of Frankfurt and the Kaiserstuhl Massif near Freiburg, famous for its calcium-rich carbonatite rocks.

The Rhineland-Palatinate

The majority of German vineyards are located on either side of the Rhine Graben, with the largest area of production in the Rhineland-Palatinate, a northward extension of the Alsace slopes. The pattern of geology and climate is much the same as in Alsace, and so is the landscape. The vineyards run in a continuous strip along the Permian-Triassic sandstone ridges of the graben, protected from the weather by the Pfalzerwald (a wooded, mountainous area), overflowing in places into the Tertiary and Quaternary fill on the plain. The vines are the same as in Alsace, with the addition of the Müller-Thurgau and the Kerner. The Riesling is increasingly popular, yielding wines with more body than those from other German regions and characteristics that are very similar to their French cousins.

Rhinehesse Palatinate

This is the northward extension of the Rhineland-Palatinate, a region of much the same area situated in the meander formed by the Rhine, as it veers sharply westward to skirt the Taunus Massif (northeast of Wiesbaden). At this point, the edge of the Paleozoic bedrock is farther from the river. The vineyards carpet the gently rolling hills, supported by Oligocene and Miocene terrain (marls, sands, and scree). The predominant grape variety, more widely planted here than anywhere else in the world, is the Silvaner, which produces very dry, crisp, and fruity wines. The most famous Rhinehesse wines come from a steep-sided valley called the Rheinterrasse, between Nackenheim and Oppenheim, where the Paleozoic bedrock outcrops at the surface. Riesling wines originating from these red, clayey schistose soils are among the most elegant in Germany, the dry wines being especially fragrant (with hints of apricot and peach). Oppenheim and Nierstein Riesling wines are rich in residual grape sugars and display a wide range of aromas with age.

Baden

The Baden vineyards extend along the right bank of the Rhine from Lake Constance to Heidelberg, on the piedmont of Triassic sandstone and Black Forest granite. Plantings are less dense here than on the western margin of the Rhine Graben and confined to sheltered positions on well-exposed slopes. The heart of wine production is the basaltic terrain around the Kaiserstuhl (a dormant volcano) and the Jurassic limestone of the Tuniberg area, opposite Freiburg. The Müller-Thurgau is the grape of choice for white wine production, yielding very perfumed wines that in good years can become quite elegant. The Sylvaner also shows to good advantage in the Kaisterstuhl basalts, where it yields wines of exceptional finesse. The Ortenau region is Riesling country, producing particularly fine wines with more delicacy than their Alsace cousins, thanks to a bracing acidity and less body. The red wines meanwhile owe everything to the Pinot Noir, which yields generous, fine, and fleshy wines, from the Ortenau area through to the Kaiserstuhl and on up to the Tuniberg region.

The Rhine Valley in the Buppard region, Rhineland-Palatinate.

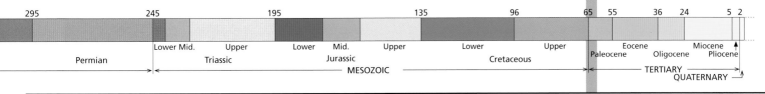

Burgundy

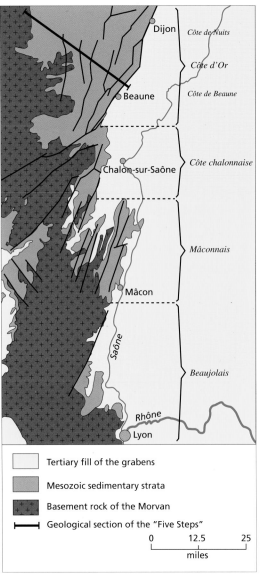

Geological overview of Burgundy and Beaujolais.

Legend for map:
- Tertiary fill of the grabens
- Mesozoic sedimentary strata
- Basement rock of the Morvan
- Geological section of the "Five Steps"

0   12.5   25
miles

The combined Burgundy and Beaujolais area represents a total planting area of 50,000 hectares. It is part of a northerly wine-growing region known as Greater Burgundy, which extends north to south in eastern France. The climate is semi-continental but subject to a wide range of influences depending on topography, altitude, and the distribution of waterways. In geological terms, Greater Burgundy consists of two distinct regions: the sedimentary basin supporting the vineyards of the department of the Yonne (*cf.* p. 74); and the Côte d'Or – or simply La Côte – which was produced, like Alsace, by successive periods of collapse and uplift.

Greater Burgundy covers a 125 square mile area and extends mainly north-south in eastern France. In simple terms, the landscape is defined by the collapse and uplift that marks a break between the Saône Plain to the east and the Morvan to the west. That movement however, did not occur along a single, large north-south fault, but rather along a very dense network of essentially south-southwest and north-northeast faults. This created a landscape characterized by east-facing slopes ideal for viticulture. Elsewhere, the ancient bedrock of the Morvan, which shows through at the surface 35 miles from Dijon, gradually draws closer to the Saône Plain and actually meets up with it in the Beaujolais region. This crustal deformation led to variations in the sequence of Mesozoic sedimentary strata covering the ancient bedrock and created four viticultural sub groups, each with a distinctive geological profile: the Côte de Nuits and the Côte de Beaune, the Côte Chalonnaise, the Mâconnais, and the Beaujolais.

The Côte de Nuits and the Côte de Beaune

The Morvan and the Saône plateau are linked by a giant stairway of five immense steps that gradually descend from west to east. They are clearly visible from the A6 freeway

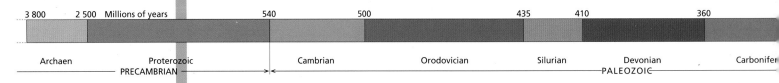

| 3 800 | 2 500 | Millions of years | 540 | 500 | 435 | 410 | 360 |

| Archaen | Proterozoic | Cambrian | Orodovician | Silurian | Devonian | Carbonifer |
| PRECAMBRIAN | PALEOZOIC |

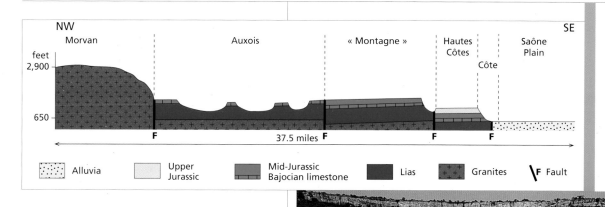

NW — Morvan — Auxois — « Montagne » — Hautes Côtes — Côte — Saône Plain — SE

feet
2,900

650

F — 37.5 miles — F — F — F

| Alluvia | Upper Jurassic | Mid-Jurassic Bajocian limestone | Lias | Granites | F Fault |

The five steps of Burgundy.

between Pouilley-en-Auxois and Beaune. The highest points, visible in the distance toward the east as you approach from the north of Pouilly, are the basement-granite crests of the *Morvan* (around 2,900 feet). The second step corresponds to the *Pouilly-en-Auxois* area, where softer Liassic marls were more rapidly eroded leaving the topographic bowl that we see today. Shortly after the fly-past over the Canal de Bourgogne, the road climbs onto the third step, a plateau area known as the *"Montagne"* (Hill of Corton) that ends at the mountain pass of Bessey-en-Chaume and borders on the Hautes-Côtes de Beaune region. Although the ground surface is higher here, the strata are sheared below the levels of the previous step. The fourth step, where the freeway runs steeply downward, is the *Hautes-Côtes* region formed from Upper Jurassic (Oxfordian) strata. Along its margins are the vineyards of the Côte de Beaune. Shortly afterward, the freeway descends the fifth and final step: the *Saône Plain*, formed from Tertiary detrital fill.

The Côte Chalonnaise and the Mâconnais

From Chagny to Givry, the Paleozoic bedrock is visible as a horst, or upthrust block, between two faults. The Mont-Saint-Vincent horst was uplifted at the same time as the Alps and only retained its sedimentary layer in the north and east, where it meets the plain. To understand what happened, slip your hand, fingers curled and palm downward, into a sandpit, until a thin layer of sand covers the back of your hand. If you lift the hand slightly, the sand will slide off the back of the hand but remain on the fingers. The hand represents the basement rock, while the sand represents the Jurassic sedimentary layers that support the vineyards.

South of Givry, at the end of the Mesozoic Era, the Côte Chalonaise from Montagny to Saint-Gengoux-le-National formed a vast anticline* with the hills of the Mâconnais (the

region around Mâcon). This collapsed in the Oligocene, creating the Grosne Rift Valley. The terrain we see today is formed of west-tilted strata in the Côte Chalonnaise and east-tilted fault blocks in the hills of the Mâconnais, with exposed basement rock beneath the Mesozoic sedimentary layers.

Beaujolais Viticultural Area

Southwest of Villefranche-sur-Saône, with the exception of a small region known as Les Pierres Dorées (the Golden Stones), the basement rock meets up with the Saône Plain, beneath the vineyards of the Beajolais-Villages Appellation and the ten following crus: Saint-Amour, Juliénas, Chénas, Moulin-à-Vent, Fleurie, Chiroubles, Morgon, Régnié, Côte-de-Brouilly and Brouilly. This geological situation has a fundamental effect on red wine production. The Pinot Noir reigns supreme from Dijon southward on Mesozoic sedimentary limestone strata, but the Gamay Noir takes over in the acidic, granite terrain of the Beaujolais.

The village of Saint-Romain straddles the fault between the third step (the Montagne) and the fourth (the Hautes-Côtes and the Côte). Visible in the background, are the steep cliffs of hard Bajocian limestone (eastern edge of the Montagne plateau) above the Liassic marls that support part of the Hautes-Côtes de Beaune (behind the village). The terrain in the foreground is from the Upper Jurassic.

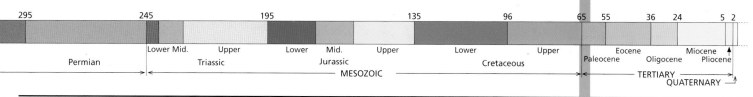

295		245		195			135		96		65	55		36	24		5 2

| Permian | | Lower Mid. | Upper | Lower | Mid. | Upper | | Lower | Upper | | Paleocene | Eocene | Oligocene | | Miocene Pliocene |

Triassic — Jurassic — Cretaceous

MESOZOIC — TERTIARY — QUATERNARY

The Côte de Nuits and the Côte de Beaune

The Côte d'Or (literally the "golden slope") – or simply La Côte – is a narrow strip of vineyard slopes that stretches for some 30 miles from Dijon to Les Maranges, renowned for such famous wine villages as Chambertin, Musigny, Vosne-Romanée, Meursault, and Montrachet. It encompasses the vineyards of the Côte de Beaune and the Côte de Nuits, on the fourth step of the vast stairway that descends from the Morvan to the Saône Plain. However homogeneous they may appear, the Côte de Nuits and the Côte de Beaune are in fact quite distinct in terms of both topography and geology.

Schematically, the great vineyards of the Côte d'Or are located on the side of the fourth step, above the plain, while those of the Hautes-Côtes occupy the more or less flat upper section. This fourth step is made up of a sequence of superimposed, alternating hard and soft strata that punctuate the landscape. Variations in strata, from top to bottom, are as follows:

- Liassic marls and clays (Toarcian);
- Bajocian *Calcaire à entroques**
  (hard crinoidal limestone);
- *Ostrea acuminata** marls (Upper Bajocian);
- Relatively soft Premeaux limestone (Lower Bathonian);
- White Oolite* (Mid-Bathonian);
- Hard, compact Comblanchien* limestone
  (Mid-Bathonian);
- Oolitic, hard *Dalle nacrée* (Upper Bathonian
  pearly flagstone) that takes its name from the rock's
  pearly luster;
- Callovian Oxfordian ferruginous oolite;
- Mid-Oxfordian, relatively soft marls and limestone;
- Hard Oxfordian limestone;
- Kimmeridgian limestone (supports no vineyards on
  the Côte d'Or);

The vineyard of the Domaine Comte de Vogüe, planted on shallow argillaceous-calcareous soils on the outskirts of the village of Chambolle-Musigny.

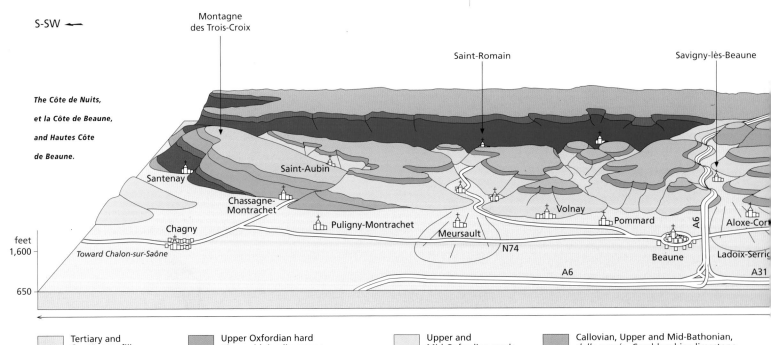

S-SW ←

The Côte de Nuits, et la Côte de Beaune, and Hautes Côte de Beaune.

Montagne des Trois-Croix

Saint-Romain

Savigny-lès-Beaune

Santenay

Saint-Aubin

Chassagne-Montrachet

Chagny

Toward Chalon-sur-Saône

Puligny-Montrachet

Meursault

N74

Volnay

Pommard

Beaune

A6

Aloxe-Cor

Ladoix-Serrig

A6

A31

feet
1,600

650

| Tertiary and Quaternary fill | Upper Oxfordian hard Kimmeridgian limestone | Upper and Mid-Oxfordian marls | Callovian, Upper and Mid-Bathonian, *dalle nacrée*, Comblanchien limestone |

The multiply-owned, 50-hectare Clos de Vougeot estate (with 85 owners at last count) is classically divided into three distinct sections: an upper section of shallow Bajocian limestone: a middle area of limestone scree (shown in the foreground); and a lower section of dark brown soils on marls.

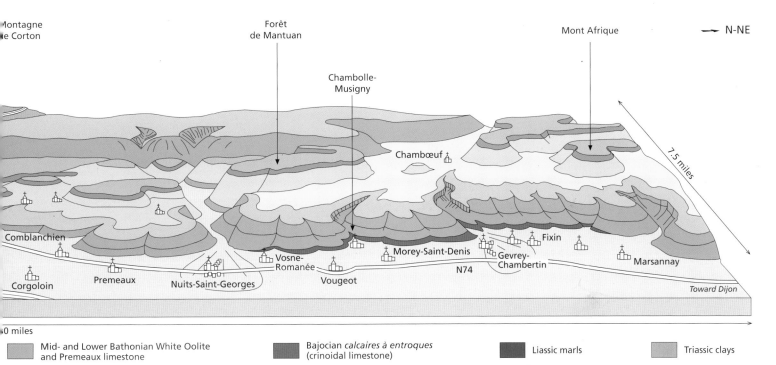

Montagne de Corton

Forêt de Mantuan

Chambolle-Musigny

Mont Afrique

→ N-NE

Chambœuf

7.5 miles

Comblanchien

Fixin

Vosne-Romanée

Morey-Saint-Denis

Gevrey-Chambertin

Marsannay

Premeaux

Nuits-Saint-Georges

Vougeot

N74

Corgoloin

Toward Dijon

0 miles

Mid- and Lower Bathonian White Oolite and Premeaux limestone	Bajocian *calcaires à entroques* (crinoidal limestone)	Liassic marls	Triassic clays

On the Côte de Nuits, the Jurassic strata gradually sink as you move south. At mid-slope in the Gevrey-Chambertin Grands Crus, the Bajocian limestones are found at the base of the slope in Vougeot, then disappear beneath the Bathonian (White Ooolite and Premeaux stone) in Vosne-Romanée.

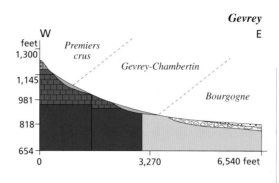

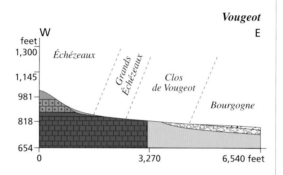

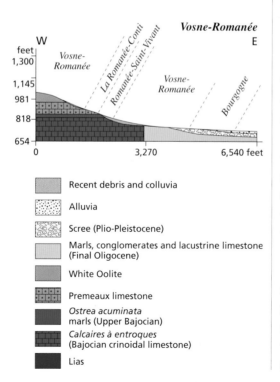

Legend:
- Recent debris and colluvia
- Alluvia
- Scree (Plio-Pleistocene)
- Marls, conglomerates and lacustrine limestone (Final Oligocene)
- White Oolite
- Premeaux limestone
- *Ostrea acuminata* marls (Upper Bajocian)
- *Calcaires à entroques* (Bajocian crinoidal limestone)
- Lias

At the Top of the Côte de Nuits: Comblanchien Limestone

The Côte de Nuits extends in Napoleon-like layers all the way from Dijon to Chagny. It is not perfectly homogenous in structure, however, and if you could see the entire Côte at a glance you would notice two broad undulations: the upward arching of the Côte de Nuits (Gevrey anticline*) and the downward warping of the Côte de Beaune (Volnay syncline*). These waves stand out in the landscape as Comblanchien and Upper Oxfordian hard limestone strata.

The cap rock overlying the vineyard slopes of the Côte de Nuit is Comblanchien limestone, rising from around 1,100 feet in Marsannay to more than 1,300 feet above Gevrey-Chambertin. It then gradually descends to some 900 feet in Comblanchien, the village from which the geological formation takes its name. Yellow, compact Comblanchien limestone is also used as a building stone by the communes of Corgoloin and Premeaux. This is the rockiest section of the Côte, known in French as the Côte des Pierres ("Rocky Côte").

At the Top of the Côte de Beaune: Oxfordian Strata

The Upper Oxfordian formation is located at more than 1,900 feet on the Mont Afrique in the Hautes Côtes de Beaune and descends slightly to 1,700 feet in the Forêt de Mantuan. The lower forest zone supports the first Hautes Côtes de Nuits vineyards, planted on Mid-Oxfordian marls. The Comblanchien disappears beneath the level of the Saône plain around Ladoix-Serrigny, while the compact Oxfordian limestone crowning the Hill of Corton rises to more than 1,200 feet. It will remain the cap rock for most of the Côte de Beaune slopes.

In Volnay, the Ancient Bedrock Resurfaces

In Volnay the movement of the strata is reversed. Comblanchien limestone re-emerges from the plain south of Meursault, rapidly rising to more than 1,600 feet on the Montagne des Trois-Croix, where the Côte de Beaune comes to an end. The Oxfordian is stripped away by erosion and stops at the small

3 800	2 500	Millions of years	540	500	435	410	360
Archaen	Proterozoic		Cambrian	Orodovician	Silurian	Devonian	Carbonife
	PRECAMBRIAN					PALEOZOIC	

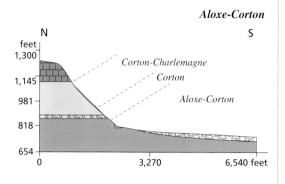

Aloxe-Corton

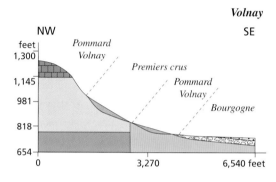

Volnay

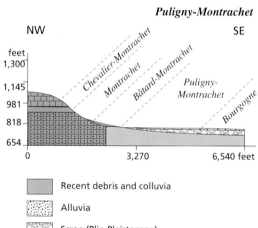

Puligny-Montrachet

Recent debris and colluvia

Alluvia

Scree (Plio-Pleistocene)

Marls, conglomerates and lacustrine limestone (final Oligocene)

Hard limestone (Upper Oxfordian)

Marls and limestones (Middle and Upper Oxfordian)

Ferruginous Oolite (Lower Oxfordian)

Dalle nacrée (pearly flagstone)

Comblanchien limestone

White Oolite

Premeaux limestone

valley of Saint-Aubain. This rapid resurgence of the strata is due to the uplift of the Morvan basement rock, which is close at hand in the Maranges area at the base of the Cosanne Valley, and around the village of Couchey.

The Forces of Erosion

The repercussions of this double undulation are apparent in the topography and therefore the vineyards of the Côte de Nuits and the Côte de Beaune. Witness the single, deep indentation in the Côte de Nuits in the Meuzin River valley, home of Nuits St. Georges, compared with the series of inlets along the small slope of the Côte de Beaune, at Savigny-lès-Beaune (Rhoin River), Pommard (Avant-Dheune River), Meursault (St.-Romain stream) and between Chassagne and Puligny-Montrachet (St. Aubin valley).

The reason is simple. On the Côte de Nuits, the top of the small slope of the Côte is protected by a tough carapace of Comblanchien limestone that has largely escaped erosion. There are just a few dry valleys, only two of which give access to the Chambœuf Plateau: the Combe* Lavaux in Gevrey-Chambertin and the Combe* in Chambolle-Musigny. As a result, all the Grands Crus, Premiers Crus, and *Villages* vineyards from Dijon to Ladoix-Serrigny are confined to a band some three-quarters of a mile wide.

The Côte de Beaune, on the other hand, consists essentially of Oxfordian marls and limestone, which have proved less resistant to erosion. This explains those deeply incised valleys that divide the Hautes Côtes de Beaune, and extend all the way to the foot of the Montagne cliff. The Grands Crus are scattered along the Côte, but the *Villages* and even Premiers Crus vineyards are planted in the valleys as far as Saint-Aubin and even Saint-Romain, close to the Montagne.

Complex Soils

Burgundian soils are rarely derived exclusively from weathering of the underlying strata. All along the Côte, these strata have been overlain by detritus of varying thickness, consisting primarily of red slope wash that slid down the plateaus and slopes

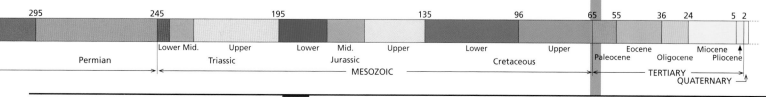

in the Pleistocene Ice Age by a process known as solifluction*. It then mixed with erosional scree, from the more or less hard limestone at the top of the Côte and from the blocks of rock upthrown by successive collapses. This hodgepodge of ingredients became mixed with the existing soil, greatly assisted by the traditional Burgundian practice of turning over the soil after heavy rain.

Today, the parent rock* often outcrops on the steepest slopes (Ruchottes-Chambertin, Clos de la Roche in Morey-St.-Denis, Les Reignots in Vosne-Romanée, and Chevalier-Montrachet) but the layer of scree can be quite thick in flatter areas (Grands-Echézeaux). This leads to marked differences in depth of rooting, resistance to drought, and average productivity.

Along the Côte de Nuits Wine Route

A trip along the Wine Route provides a detailed picture of the development of the Côte d'Or terroirs. The journey starts with the Fixin and Marsannay vineyards planted in Bajocian limestone and marls. After Brochon, the Gevrey-Chambertin appellation boasts a range of terroirs as you leave Combe Lavaux. To the north, the Premiers Crus are planted on Bajocian *Ostrea acuminata* marls mixed with red alluvia and slope wash from the plateau. Beneath Gevrey-Chambertin, a large alluvial fan from Combe Lavaux allows the *Villages* appellation to extend nearly to the railway. South of the village, the Chambertin Grands Crus are supported by hard Bajocian crinoidal limestone. Here shallow soils combined with an east-facing exposure bring color, body, finesse and strength to the Chambertin red wines. Conditions remain the same in Morey-St.-Denis and in the northern part of Chambolle-Musigny. The crinoidal limestone then falls away to the upper part of the Clos de Vougeot where the soils are shallow, while the central section consists of limestone scree, and the lower section of deep brown soils over marls. The crinoidal limestone disappears altogether beneath the village of Vosne-Romanée.

Premeaux limestone and White Oolite take over in the upper sections of the Richebourg, Romanée, Romanée-Conti, La Grande Rue and La Tâche Grands Crus, all renowned for their long-lived wines. The narrower Nuits-St.-Georges vineyard also boasts some first-class Premiers Crus supported by Premeaux limestone.

The Hill of Corton (or Montagne) and the Côte de Beaune

Ladoix-Serrigny, dominated by the imposing Hill of Corton, marks a dramatic change in landscape as well as the start of the Côte de Beaune. The Montagne is the epitome of the Côte d'Or's stratigraphic character, from the Oxfordian cap rock to the compact limestone base. The pyramid-shaped, steeply sloping Montagne extends northeast-southeast on a solid foundation of tough limestone strata (compact *dalle nacrée*, overlaid by ferruginous oolite, then by limestone mixed with sandstone and marl). Halfway up the slope, the limestone group is overlain by Mid-Oxfordian marls and the wooded top of the hill is capped by Upper Oxfordian Oolitic* and Bioclastic* limestone. The foot of the hill is the domain of choice for Pinot Noir that in these brown, shallow limestone soils yields robust, balanced, velvety red wines (Aloxe-Corton AOC). The Oxfordian marls further up the slope are occupied by the Chardonnay, source of the distinguished white wines of Corton-Charlemagne that are renowned for their generosity and impressive longevity. The geological sequence remains unchanged in the neighboring villages of Pernand-Vergelesses and Savigny-lès-Beaune where there is the same affinity between soil and grapes. There is a change in the red wines however, that are less robust here than those from the Aloxe-Corton side while developing notes of cherry that grow more pronounced the closer you get to Savigny-lès-Beaune. There are also widespread vineyards at the foot of these slopes, planted on flat alluvial terrain, rich with limestone slope wash from the Rhoin River and the Pernand Valley. Farther south on the Côte, the vineyards of Beaune, Pommard, and Volnay are established on Oxfordian marls and limestones, mixed with reddish clays and slope wash. Pinot Noir reigns supreme here, producing the powerful, structured wines of Pommard and the rather more elegant Volnay wines.

The Meursault commune is a transitional terroir. The northern part is the continuation of the Volnay slopes, while the

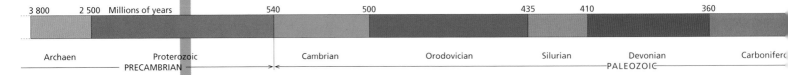

southern part marks the resurgence of Comblanchien limestone, more dolomitic* here than in the Côte de Nuits. The Pinot gives way to the Chardonnay, which monopolizes the vineyards of Puligny and Chassagne. These slopes of Mid- and Lower Bathonian shallow rendzina* soils are home to the Montrachet, a nearly perfect white wine. The distinctly stony, steeply inclined terrain around Chevalier-Montrachet grows more gentle in Montrachet, becoming almost flat and richer in alluvia and clays in the lower part of Bâtard-Montrachet. The vineyards here are priceless. Toward the south, the Bathonian limestone continues to rise above Santenay (another dedicated red wine area, famous for its fleshy, powerful wines) eventually capping the slopes of Les Maranges.

Dezize-Lès-Maranges: The Test of Geology

No other French wine-growing region boasts as many appellations as Burgundy. But is there really a link between the geological formations that support the vines and the characteristics of Burgundian wines? It's difficult to answer this question, since the greatest Grands Crus are distributed throughout the Jurassic formations: Bajocian limestone in the Chambertin AOC, Premeaux limestone and White Oolite in Romanée-Conti, the Comblanchien in Montrachet, and Oxfordian marls in Corton. On the other hand, no one could deny the influence of geology in shaping the hierarchy of wines that runs from the southern tip of the Côte de Beaune in the area of Les Maranges.

The little village of Dezize-Lès-Maranges clings to the sheer southern face of the plateau crowned by the Montagne des Trois-Croix. On either side of the village are two vine-carpeted slopes. Seen from the other side of the Cosane River Valley, there is no obvious difference between the plateau and the slopes. And yet the large fault that separates the Côte and the Hautes-Côtes zone, to the east, from that of the Montagne, to the west, passes through the center of the village. The edge of the plateau only appears uniform because the Sinemurian limestone (Liassic), to the west, and the Bajocian limestone, to the east, have been brought to the same level by this fault. In the eastern part, the Sinemurian is located 260 feet farther down, at the foot of the slope.

The vine found the conditions that suited it best. There is a small corrie overlooking the village that is classified as a *Villages* appellation. But otherwise, in viticultural terms, the difference between the two slopes is striking. Witness the difference in the wines: to the west, Bourgogne Appellation Régionale (Triassic clays) and Bourgogne AOC Grand Ordinaire (Triassic sandstone and granites at the base of the slope); to the east, exclusively Maranges Premiers Crus (Liassic marls).

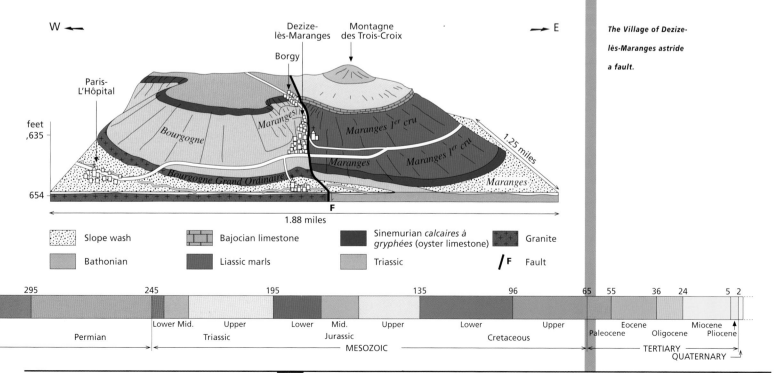

feet
,635

654

1.88 miles

| Slope wash | Bajocian limestone | Sinemurian *calcaires à gryphées* (oyster limestone) | Granite |
| Bathonian | Liassic marls | Triassic | /F Fault |

The Village of Dezize-lès-Maranges astride a fault.

295   245   195   135   96   65   55   36   24   5 2

Permian | Lower Mid. Upper — Triassic | Lower Mid. Upper — Jurassic | Lower Upper — Cretaceous | Paleocene | Eocene Oligocene | Miocene Pliocene

MESOZOIC — TERTIARY — QUATERNARY

The Pinot Noir

The Pinot Noir is the most unique and certainly the most capricious of all the great grape varieties. It is also the most sensitive to the infinite variations in soil and has come to symbolize that almost mythical link between grape and terroir. Best of all for the home-loving Pinot Noir is its native Burgundy. Unlike most other grapes that have successfully adapted to a variety of French wine-growing regions, the Pinot Noir in Burgundy produces results that are unequalled elsewhere.

Let us now look at the pattern of Pinot Noir plantings along the Saône graben from Dijon toward Lyon. Note that the Pinot Noir is extremely choosy and will only settle for Jurassic limestone formations. It has an aversion to acidic terrain and avoids those areas where the Jurassic formations have risen up to expose the subjacent Hercynian bedrock at the southern end of the Côte de Beaune. Witness the absence of Pinot Noir in the Couchey area, on soils formed from the acidic granites of the Morvan and the equally acidic Triassic sandstone foundation. This is where the Gamay takes over. The rivalry between the two grapes continues in the Mâcon area. Here the Hercynian basement is sliced into parallel strips overlaid by Jurassic limestone strata, leaving outcrops of granitic terrain at their base that discourage the Pinot Noir. It then disappears altogether from the Beaujolais area, where the granitic sands and schists of the Hercynian basement meet up with the Saône plain.

In the Pommard appellation, the Pinot Noir grows in deep, argillaceous soils formed from Oxfordian marls, which yield powerfully structured, buxom wines with rich, ripe fruit aromas.

A Love of Limestone

In limestone terrain, the Pinot Noir can inspire an infinite range of wines, including the magnificent Romanée-Conti. The basis of the Pinot Noir hierarchy remains a mystery, however, as no one has ever managed to decode the language of the terroir. Its love of limestone extends to every type of Jurassic formation, ranging from Liassic marls in the Maranges Premier Cru; to Bajocian crinoidal limestone in Gevrey-Chambertin and the Clos de Vougeot; to Premeaux limestone in Vosne-Romanée; and Callovian-Oxfordian marls and limestones in Pommard and Volnay. In 1981, a team of researchers led by Suzanne Mériaux demonstrated that the Grands Crus are always located on the stoniest, most limestone soils, usually on medium slopes. But there is much more to it than that.

The Burgundian Expressions of the Pinot Noir

Travelling from Dijon to Chagny, we first come across the Pinot Noir in the Marsannay AOC where it is the unexpected source of discretely floral rosé wines with notes of fresh grapes. These days, Marsannay and the neighboring Fixin appellation concentrate on red wine production. Here, Pinot Noir is grown in soils formed from Bajocian crinoidal limestone. The results are red wines with aromas of violets, black currant and blackberry, and a solidly structured palate, thanks to robust tannins that need to mellow.

The Pinot Noir grows in the same soils in the Gevrey-Chambertin AOC and, more particularly, the Chambertin Grands Crus. The results here are fleshy, strong, corpulent wines that take time to mature. Their range of aromas is enriched by notes of cherries in eau-de-vie, plus scents of game and leather as the wine ages. The Griotte-Chambertin owes its name not to morello cherries (**griotte**), as some might think, but to the pebbly terrain known locally as crais*. Morey-St.-Denis also has a reputation for solidly structured wines. The wines of Chambolle-Musigny, in contrast, are subtler and more elegant. Its vineyards are supported by Bajocian limestone formations that drop away toward the village exposing marls capped by Premeaux limestone on the slopes. But nobody can say for sure whether this change in geology is responsible for the altogether more feminine appeal of a Chambolle-Musigny. It could also be due to the influence of the Combe-Chambolle, or some other factor entirely.

The Bajocian strata start to slope in the neighboring Clos de Vougeot, where the wines have more vigor but the same elegance and class. The Vosne-Romanée Grands Crus are established on the marls and Premeaux limestone that caps the Bajocian formation. They include the vineyards of Richebourg and Romanée, where the Pinot Noir achieves optimum elegance and aromatic complexity.

The strata continue to dip beneath the level of the Saône plain around Nuits-St.-Georges, home to velvety wines with palpable tannic structure. In the Hill of Corton, the Pinot Noir grows in slabby limestone known as dalle nacrée* that is overlaid at the base of the slope by ferruginous oolite. These crus yield chewy, full-bodied wines that tend to be somewhat tannic in their early years. Wines from neighboring communes supported by the same formations are less powerful but still show a family resemblance; the Savigny-lès-Beaune Pinot Noirs have an added nuance of cherries in eau-de-vie.

After sharing space with the Chardonnay from Nuits-St.-Georges onward, the Pinot Noir once again reigns in Pommard and Volnay, where it thrives on more recent Mid- and Upper Oxfordian strata. However similar their geological foundations, the two communes produce quite different wines: Pommard is fleshy and strong with aromas of ripe fruits; Volnay is feminine and subtle with crisp, fresh notes. There are, in fact, small but important distinctions between the terroirs. Pommard vineyards are south- and southeast-facing, and the soils are deeper and more clayey. Volnay vineyards tend to be more east-facing, and the soils are leaner but richer in limestone.

As the geological formations gradually rise up again toward the south, the Pinot Noir gives way to the Chardonnay.

It then makes a comeback in Santenay in an extremely complex terroir broken up by a patchwork of faulting. West of Santenay, Les Maranges boasts the oldest strata on the entire Côte d'Or: Liassic marls that were barely noticeable in Brochon to the

The Marsannay-la-Côte appellation, where the Pinot Noir produces fleshy, fruity rosés that are unique on the Côte de Nuits. Brown limestone soils rich with pebbles and gravel provide good drainage.

north. These colder soils yield more rugged, tannic wines that in youth leave a touch of bitterness on the palate.

In the Côte Chalonnaise, there is a return to the Upper Jurassic formations previously found in Pommard and Volnay. The strata here dive down toward the Saône Valley. The Pinot Noir is back in familiar territory, yielding full, honest, fleshy wines in Rully; rounded, perfumed wines in Givry; and well-built, elegant wines in Mercurey. The wines from this appellation may be powerful or supple, depending on whether they originate in limestone or argillaceous soils. The Chardonnay then takes over in the Liassic marls of Montagny. The Pinot Noir returns in the shallow limestone soils of the Mâconnais area where the Gamay makes its debut.

Northern Vineyards: The Accent on Finesse

The Pinot Noir only does well in cool, northerly climates. It is one of the authorized grape varieties in Champagne (northern France), where it grows mainly in the chalk surrounding the Montagne de Reims. The bulk of production is vinified as white wine for making into Champagne. There is also a red, non-sparkling AOC Coteaux-Champenois made from Pinot Noir, the most famous being Bouzy red. Pinot Noir wines from the Kimmeridgian soils of the Aube or the Sancerre region display unmistakable aromas of ripe morello cherries.

Farther north, the Pinot Noir grows in the marls on the borders of the Jura, and in the limestone soils of the sub-Vosgian Hills in Alsace. These wines are paler, with a lighter structure, and offer a foretaste of wines from the German side of the Rhine River. The Pinot Noir is the premier red grape variety in Germany, although it represents only five per cent of total planting acreage.

Across the Atlantic: Oregon Stands Out

American winegrowers have been much baffled by the Pinot Noir, but it has proved the making of the Los Carneros AVA (American Viticultural Area), located in Napa and Sonoma counties. This is the cooler end of the Napa Valley, fanned by breezes from San Francisco Bay and shielded by the Mayacamas Mountains. The Pinot Noir also does well in the Russian River Valley in Sonoma County. But it really comes into its own in the state of Oregon, in the Willamette Valley south of Portland. Here basaltic soils combined with cooler weather conditions bring out the very best in the Pinot Noir. Full-bodied, fruity, and well structured, the wines of Willamette Valley are the finest expression of Pinot Noir outside France.

Volnay and Pommard wines both originate in Oxfordian marly slopes, but they show the Pinot Noir in a quite different light: delicate in Volnay, robust in Pommard.

Vosne-Romanée: the Pinot Noir planted on gently sloping Premeaux limestone and white oolite surpasses itself in the wines of Romanée and Romanée-Conti.

The Chardonnay

The Chardonnay and the Pinot Noir, on home territory in Burgundy, are one of the most famous "wine couples" in the world. They are far from well matched however, each having quite distinctive characteristics. The whimsical Pinot Noir is only happy in its native Burgundy, and, as many a despairing winegrower knows, refuses to adapt to vineyards beyond sight of the Saône. The seductive, endlessly charming Chardonnay, however, is happy wherever it goes and, unlike the Pinot Noir, has invaded vineyards all around the world.

On the Hill of Corton, the Chardonnay lines steep south- and southwest-facing slopes of Argovian marls.

But the Chardonnay is plainly at its most dazzling in Burgundy. Although cultivated in Champagne and Alsace for the production of effervescent wine, the Chardonnay is not restricted to northerly French vineyards. There are plantings of Chardonnay in the Jura, Savoie, the Bugey area, the Rhône Valley (Dios, Ardèche), and the Languedoc-Roussillon (Vins de Pays and Limoux) – everywhere in fact, except southwest and central France. Outside France, the Chardonnay has adapted to a wide range of vineyards, producing wines that for millions of consumers achieve a quality close to the inimitable Montrachet. In Europe, the Chardonnay been adopted by the Italians and the Spaniards but its most fervent admirers are those relative newcomers to wine-growing in the New World. American viticulturalists are particularly keen on the Chardonnay but so too are the Australians, the South Africans, the Chileans, the New Zealanders, and many others.

The Burgundian Cradle

In Burgundy, the Chardonnay prefers rather heavy, deep, argillaceous soils, although there is more to it than that. It gives of its best on the Côte de Beaune, between Puligny and Chassagne, especially in the Montrachet terroir. The hard, dry upper layer of the Côte de Nuits is formed by the Comblanchien* limestone that sinks beneath the Côte de Beaune, at its northern end, and then re-emerges here as less compact dolomitic* rock. The Puligny-Montrachet terroir is slightly richer in limestone with soils as pebbly as those of Chevalier-Montrachet. By contrast, the soils of Chassagne-Montrachet and Bâtard-Montrachet are more argillaceous and deeper. The communes of Puligny and Chassagne share the Montrachet Grand Cru but there are subtle differences between their Chardonnay wines. A Puligny-Montrachet is characterized by flinty, mineral notes plus the merest suggestion of flowers and white fruits. The palate is well structured and long-lived. A Chassagne-Montrachet, on the other hand, is dense and meaty, with warm, exuberant aromas of honey, acacia, and ripe fruits, and a full-bodied and generous palate. In Meursault, the Bathonian formations in the southern part of the appellation are capped in the north by the Combe

The Casablanca Valley in Chile, exposed to ocean influences, owes its reputation to the Chardonnay. While most Chilean vineyards are planted on flat land, this winery chose to explore granite slopes.

d'Auxey-Duresses and Callovian oolitic limestones (*dalle nacrée*). Here the Chardonnay yields powerful, fragrant wines with a bouquet of honey, beeswax, ripe fruits, and butter matched by a dense, mellow palate.

Further northward, the Chardonnay changes expression again on the Hill of Corton, where it prefers the upper section formed by Upper Jurassic Argovian marls. For reasons of geology or climate, it is planted on south- and southwest-facing slopes while the Pinot Noir climbs all the way to the top of the hill, on the east-facing slope. These wines are unrivalled for strength and built to last. A Corton-Charlemagne Chardonnay is warm and opulent, with aromas of baked apples in butter, honey, and cinnamon and a powerful, mouth-filling palate. It has the best structure of all the Burgundian crus and survives years of cellaring. The Chardonnay grows scarcer toward the north, except in Musigny and the *Villages* appellations. Southward, plantings of Chardonnay continue on the Côte Chalonnaise and in the Mâcon area. In fact, it accounts for two-thirds of planting acreage in Rully, mainly grown in Argovian marl and limestone soils that extend to Mercurey and Givry. The two other communal appellations have only minor plantings of Chardonnay. These Chardonnay wines are elegantly expressive: straightforward, fruity, and lively with a certain richness, depending on the terroir.

Everything changes after Givry, south of the geological deformation in St.-Désert. The Chardonnay reigns supreme in Montagny on argillaceous Liassic and Triassic marls, to which the wines owe their bouquet of white flowers and fruits and a pleasing, delicate palate. The Mâcon area is also dominated by the Chardonnay, producing powerful wines in shallow limestone soils, and lighter wines in argillaceous or sandy soils. West of Mâcon, the Chardonnay is once again in its element in the vineyards of Pouilly-Fuissé, Pouilly-Loché, and Pouilly-Vinzelles. The wines here acquire more complex aromas (mineral notes, touches of citrus and honey) and leave an impression of opulence on the palate.

The style is altogether different at the northern limit of Burgundy, where the chameleon-like Chardonnay again demonstrates its versatility. The vineyard in question is of course Chablis, on the A6 Autoroute from Mâcon to Paris. The remarkable Chablis terroir, embodied in the belt of Grands Crus to the north of the town of Chablis, brings out unique characteristics in the Chardonnay. Gone are the buttery and honeyed aromas of the Côte d'Or. The wines of Chablis suggest white fruits mingled with mineral notes of flint and pierre à fusil (gun flint). Their elegant palate comes from a characteristic freshness that harmonizes perfectly with the wine's plump, pleasing substance.

In the southern part of the Côte Chalonnaise, thanks to the reversing of the slope of the Jurassic strata, the Chardonnay enjoys a sheltered location on the Liassic marly slopes of the Montagny AOC (pictured here, the village of St.-Vallerin).

freshness that harmonizes perfectly with the wine's plump, pleasing substance.

Aube Chardonnay

The Chardonnay is very much at home in the Aube, a Champagne sub-district located on the Kimmeridgian rim of the Paris Basin. Note however that the best Blancs de Blancs Champagnes come from the department of Champagne itself, notably from the Kimmeridgian strata closer to the center of the Paris Basin, south of Epernay. This is where we find the Côte des Blancs, a scarp with Eocene rock at the top and out-crops of Upper Cretaceous Champagne chalk along its flanks. Côte de Blancs Chardonnay yields the finest, freshest Champagnes of all, with lemony aromas of white fruits, toast and brioche and supremely elegant effervescence.

The New Faces of the Chardonnay

In the rest of France, the Chardonnay tends to be blended with other local grape varieties. For instance, Crémant-d'Alsace is a blend of the Chardonnay and Alsace cultivars; the sweet, yellow *Vins Jaunes* from the Jura are made from the Savagnin and the Chardonnay. It does however excel in one particular region: the Limoux, where it was introduced in 1960 for the making of Blanquette-de-Limoux. It also yields a remarkably good dry AOC Limoux. These days, the Limoux Wine Cooperative produces four cuvées, each with that un-mistakable Chardonnay signature but revealing four distinc-tive personalities from various terroirs: Mediterranean, ocea-nic, the Autan terroir or the Haute Valley terroir (*cf.* Limoux, p. 110).

The story is fairly similar on the other side of the globe, in Chile. The Casablanca Valley, some 38 miles west of Santiago on the road to Valparaiso, welcomed its first vineyards in the 1980s. The climate here is exposed to the Pacific Ocean and therefore cooler than in the central part of the valley, which is shielded from maritime influences. The Chardonnay has found its niche in these fresher, granitic, arenaceous soils, pro-ducing some of the best white wines in the whole of Chile. It enjoys a similar location in California, in the Carneros region

The Côte Chalonnaise

The gently rolling vineyards of the Côte Chalonnaise cover some 4,500 hectares between Chagny and St. Gengoux-le-National. In the middle is Chalon, an ancient trading port on the Saône River. The Côte is a narrow, faulted ridge where the vines enjoy various exposures although most of the slopes face southeast. The Pinot Noir and the Chardonnay now grow alongside the Gamay Noir and the Aligoté. The Côte Chalonnaise is distinguished by four areas of communal appellation created by the Mont-St.-Vincent horst: Rully, Mercurey, Givry, and Montagny.

The Mont-St.-Vincent Horst

From Chagny to St.-Désert, the vineyards of the Côte Chalonnaise are influenced by the Mont-St.-Vincent horst*, a huge block of basement rock that was uplifted by an aftershock of the Alpine folding between the coal basin of Montceau-les-Mines and the Saône plain. The Jurassic sedimentary layer has been largely worn away by erosion and only survives today on the northern and eastern flanks. The vines here are in their element. The northern tip of the block that dives beneath Chagny is broken up by southwest- and northeast-facing faults, into compartments that show greater and greater subsidence toward the east:

• The Hill of the Roman camp of Chassey (1,475 feet);

• The Montagne de l'Ermitage (1,364 feet);

The vines of the Château de Rully are planted on Oxfordian argillaceous limestone soils.

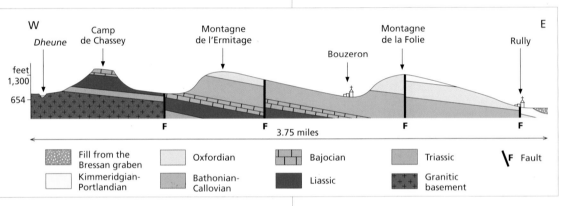

The northern tip of the Mont-Saint-Vincent horst.

The tabular relief of the Côte d'Or gives way to more rugged topography on the Côte Chalonnaise which consists of two fundamentally different geological structures, each with a pronounced effect on the vineyards. In fact, the slope of the geological layers is suddenly reversed when you reach the St.-Désert fault. This dramatic about-turn is clearly visible from the RN 80 between Chalon-sur-Saône and Montceau-les-Mines. The strata to the north slope toward the plain, while those to the south dip toward the Morvan.

• The Montagne de la Folie, with two areas of subsidence (each side of a fault running along the summit), separates the two vineyards of Rully and Bouzeron. Bouzeron is the only communal appellation to concentrate on the Aligoté, reaping superb results in vineyards that are sited on Upper and Mid- Jurassic formations, either side of a valley that gives onto Chagny.

Rully

The Rully vineyards occupy a sheltered location on the eastern side of the Montagne de la Folie, supported by Mid- and Upper Jurassic strata that gradually slope down toward the Saône Plain. The best Premiers Crus are planted on the steepest slopes,

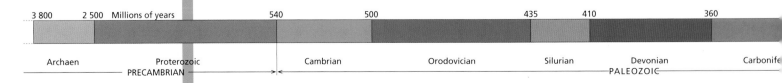

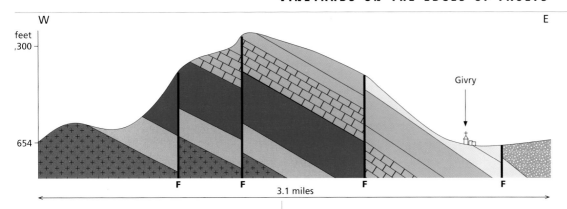

W   E

feet
,300

654

F   F   F   F

3.1 miles

Givry

*Section of the Hill
of Givry.*

on either side of the road leading from the village to the top
of the Hill of Bouzeron. The Pinot Noir yields fleshy red wines,
while the Chardonnay produces lively, aromatic, whites of
character, recalling the Côte de Beaune crus.

Mercurey

Mercurey is the continuation of the Rully appellation and
supported by the same formations. It extends into the
communes of Mercurey and St.-Martin-sous-Montaigu, where
the topography is more varied. Production is heavily dominated
by the Pinot Noir, which yields elegant, well-built reds, although
they lack the character of a Rully. The St.-Martin-sous-Montaigu
vineyards rest on Liassic formations exposed by faulting.

Givry

Matters are more straightforward in the Givry appellation area
farther south, on the edge of the plain. The Upper Jurassic
strata dip steadily eastward and the heart of the vineyard rests
on Oxfordian marly limestone soils, where the Pinot Noir yields
solid, highly perfumed reds. The landscape changes south of
St.-Désert, both in terms of geology and viticulture. The Cuesta
de Montagny in Culles-les-Roches is the western relic of a vast
collapsed arch. The Mesozoic layers capping the basement rock
are now inclined toward the west, so offering their steep edge
and no longer their sloping back to the vineyards of the
Côte Chalonnaise. The ancient granitic basement, together with its Triassic sandstone
covering, outcrops along the edge of the plain
on the Hill of Bissey-sous-Cruchaud. This area
is only entitled to the Bourgogne-Grand
Ordinaire appellation, not the Bourgogne-
Côte-Chalonnaise appellation.

Montagny

The Liassic and argillaceous Triassic marls of the Montagny
appellation area have a wooded cap of compact Bajocian
crinoidal limestone that marks the top of the Côte. The pebbly
soils and hard limestone strata of Mercurey and Givry are
replaced by deep brown marly and marly sandstone soils more
suitable to the Chardonnay, which takes over here from the
Pinot Noir. Montagny Chardonnay wines are delicate, discrete,
and somewhat dry and offer a foretaste of Mâconnais wines.
The Montagny appellation extends into four communes:
Montagny (perched on crinoidal limestone); Buxy; Jully; and
St.-Vallérin (built on Sinemurian oyster limestone). It stops at
the D 981 road from Buxy to St.-Boil, where the Upper Jurassic
(Kimmeridgian) around the Grosne Valley has been lowered
along a fault that follows the line of the road. These alluvium-
covered soils are home to the Bourgogne Appellation
Régionale and Bourgogne-Grand Ordinaire wines.
South of Culles-les-Roches the vineyards are
scattered over a few, well-exposed slopes
facing the nascent hills of the
Mâconnais to the east.

*The Côte Chalonnaise:
north of St.-Désert
(Givry) the strata dip
eastward, following the
slopes of the vineyard.
To the south (Montagny)
they dip westward,
with their edges on
the vineyard side.*

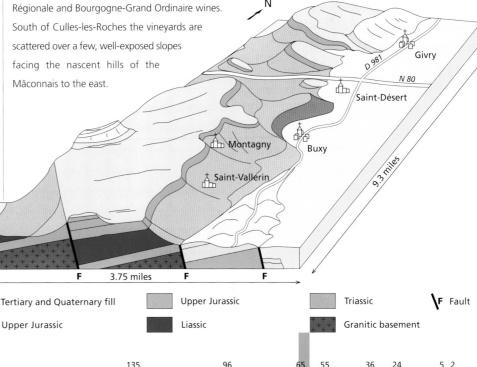

N

Givry

D 981

N 80

Saint-Désert

Montagny

Buxy

Saint-Vallérin

S

feet
1,300

654

F   3.75 miles   F   F

9.3 miles

▢ Tertiary and Quaternary fill	▢ Upper Jurassic	▢ Triassic	\F Fault
▢ Upper Jurassic	▢ Liassic	▢ Granitic basement	

295	245	195	135	96	65	55	36	24	5	2

Permian		Lower Mid.	Upper	Lower	Mid.	Upper	Lower	Upper		Eocene		Miocene
		Triassic			Jurassic		Cretaceous		Paleocene		Oligocene	Pliocene

MESOZOIC   TERTIARY
QUATERNARY

The Mâconnais

The Mâconnais, at the southern tip of the Burgundy wine region, extends north-south for about 30 miles from Tournus to Mâcon, following the peaceful Saône River. The vineyards carpet the hills up to heights of 1,300 feet, bordered to the north and west by the Grosne River, to the east by the Bresse and to the south by the Beaujolais Massif that marks the beginning of Gamay Noir country.

Chapaize chain, barely emerged from the Tertiary and Quaternary formations of the Grosne Valley.
• The second is the Blanot chain northeast of Cluny.
• The third and largest chain extends from Sennecey-le-Grand in the north to Pouilly-Fuissé in the south and features the entire series of geological formations. To the west, hilltops of basement granite protect the Mâcon region from westerly winds (La Mère Boitier 2,479 feet, the Mont de Mandé 1,991 feet and the Mont St. Romain 1,893 feet). The Triassic-Liassic

The tilted hills of the Mâconnais.

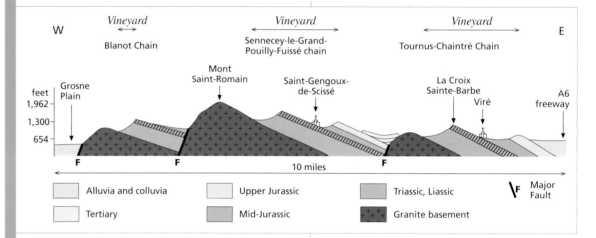

The hills of the Mâcon region represent the eastern section of a vast anticlinal arch that included the south of the Côte Chalonaise (Montagne) before it collapsed in the middle. Here the strata dip toward the Saône Plain, forming several monoclinal or non-folded blocks. Erosional forces wore away at this arrangement, accentuating the hardest parts and carving furrows through the softer, less resistant rocks. The vineyards in these small valleys occupy sheltered locations on parallel, south-southwest and north-northeast-facing slopes.

Six Compartments

The hills of the Mâconnais are compartmentalized by numerous faults but may be divided into six secondary chains that reflect practically the entire geological series.
• Leaving the Grosne Valley, the first is the Taizé-Cormatin-

depression at the foot of these hills is overlooked to the east by a Bajocian cuesta. This feature is particularly upthrown and prominent in the south where it forms the skeleton of the prow-like rocks of Solutré and Vergisson. It also supports the village of Brancion. North of the village, where the basement granite practically disappears beneath the Tertiary formations, the cuesta directly overlooks the alluvial plain of the Grosne. Above the Bajocian, a long valley formed from Oxfordian-Callovian marls is home to the villages of Cruzille, Bissy-la-Mâconnaise, Azé, Igé and Verzé. It is bordered to the east by the softer outlines of Upper Jurassic limestones.
• The fourth chain extends from Tournus to Chaintré, southwest of Mâcon. The basement granite has largely disappeared, except in Burgy northwest of Viré and west of Chaintré. The Bajocian cuesta constitutes the main landform west of Viré,

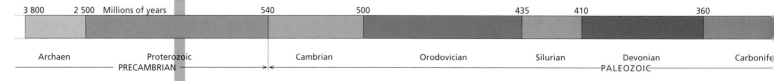

Clessé and Laizé while the Mid- and Upper Jurassic series creates a slight waviness in the landscape further east.

• A fifth, small chain consisting exclusively of the Jurassic is more developed north of Lugny.

• The last chain is a series of little hills that run along the A6 freeway from the north of Mâcon, separating the small valley of Viré-Clessé-Laizé from the Saône plain. Here once again, the only outcropping is Jurassic. Generally speaking, bearing in mind the tilt of the strata, the west-facing slopes are short and steep while the east-facing slopes are longer and gentler. Due to weathering however, the easternmost furrows have become filled with Tertiary formations, some of them quite thick, in a synclinal* arrangement that accounts for the soft, damp ground characteristic of the area. This is well illustrated by the syncline of St.-Maurice-de-Satonnay, east of Laizé. Because these slopes are relatively shallow, the Mâconnais landscape is marked by a proliferation of superficial landforms that are created by the weathering of Jurassic formations. These deposits are essentially found in areas of flat land where they mask the geological layers.

The Mâconnais Reveals Its Geology

The Mâconnais mainly produces white wines and is principally given over to the Chardonnay that does just as well on these limestone soils as the Pinot Noir. The Chardonnay thrives on rendzinas* and brown calcareous or calcic* soils where it yields wines for laying down. In siliceous, argillaceous or sandy terrain—acidic soils at the base of the soil series—it yields more easily quaffable whites for earlier drinking. The Gamay that takes over in nearby Beaujolais yields wines with great verve in these acid soils. The Chardonnay, in contrast, is the only grape to adapt to flat, reasonably dry decarbonated formations. There are five communal appellations in the Mâconnais, the most famous being Pouilly-Fuissé and its two "cousins" Pouilly-Loché and Pouilly-Vinzelles. They sit astride the southern part of the two largest secondary chains, so covering the entire geological series twice over: the Liassic at the foot of the Rock of Solutré; then the Mid- and Upper Jurassic on its eastern side;

then the Bajocian slope once again in Chaintré. St.-Véran lies on both sides of Pouilly-Fuissé, to the north and south, at the edge of the Beaujolais region, on identical formations. Farther north, at the heart of the main Tournus-Chaintré range, the vineyards of Viré-Clessé extend along the eastern side of the Bajocian slope and its overlaying formations (Upper Jurassic).

The Chardonnay wines from these appellations share distinctive characteristics although with subtle differences. Pouilly wines tend to be rich and opulent, with mineral aromas and notes of citrus mixed with accents of honey and buttery brioche. Viré-Clessé wines are fresher and livelier, with a floral bouquet (hawthorn, acacia, honeysuckle) underscored by scents of almonds and quince jam. The wines of St.-Véran lie somewhere in between.

South of the Mâconnais, the small Arlois River flows along an east-west fault that marks the end of the Mesozoic formations and the beginning of the basement granite that is the main support for the vineyards of the Beaujolais.

Colluvia and alluvia from Saône Plateau

Tertiary

Upper Jurassic

Mid-Jurassic

Triassic, Liassic

Granite basement

—— Exposed fault

---- Concealed fault

Geological section of the "tilted hills of the Mâconnais"

0   3   6.25
miles

The hills of the Mâconnais.

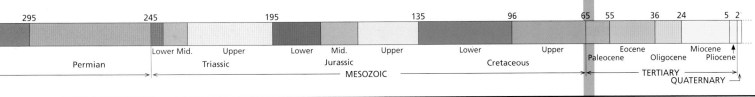

295		245		195		135		96		65	55		36	24		5	2

Permian	Lower	Mid.	Upper	Lower	Mid.	Upper	Lower	Upper	Paleocene	Eocene	Oligocene	Miocene	Pliocene
	Triassic			Jurassic			Cretaceous						

MESOZOIC — TERTIARY — QUATERNARY

The Beaujolais Region

The Beaujolais forms the eastern border of the Massif Central and represents the southern part of the Burgundy wine-growing area, to which it was officially attached in 1930. Thirty miles long and about nine miles wide, it extends from the south of the Mâconnais to the outskirts of Lyon. The rivers Mauvaise, Ardières, Vauxonne and Azergues join up with the Saône in the east, opening up valleys in the landscape. This is the land of the Gamay, which gives different results depending on whether it is grown in the northern Beaujolais, characterized by ancient granites and schistose rocks, or the southern Beaujolais, known as *Les Pierres Dorées* (the Golden Stones).

The village of Chiroubles, dominated by sloping granitic soils.

Mont Saint-Rigaud 3,299 feet

Saint-Amour
Beaujolais-villages
Chénas
Moulin-à-Vent
La Chapelle-de-Guinchay
Chiroubles
Fleurie
Morgon
Beaujolais-villages
Régnié
Ardières
Saône
Brouilly et côtes-de-Brouilly
835 m
Belleville
Beaujolais-villages
871 m
Denice
Villefranche-sur-Saône
Theizé
Azergues

Quaternary alluvia		Granites	
Colluvia		Schists	
Jurassic		Volcanic-sedimentary tuff	
Triassic			
Geological section		0 3 6.25 miles	

The Beaujolais region was uplifted higher than the Mâconnais. As seen from the A6 freeway to the south, while the hills of the Mâconnais rise no higher than 1,960 feet, the land climbs steeply south of Mâcon, reaching heights of more than 3,200 feet on the Mont St.-Rigaud, at the bottom of the Ardières Valley. The development of the lush Beaujolais slopes is linked to the collapse of the Bresse Graben in the Oligocene and to the uplift of the Beaujolais Hills during the Alpine formation. What determines the nature of the terroirs however is the Hercynian basement and its granite, metamorphic and volcanic formations. The Gamay, that "infamous plant" banned from limestone Burgundy by Philip the Bold in the 14th century, is in its element in this acidic terrain.

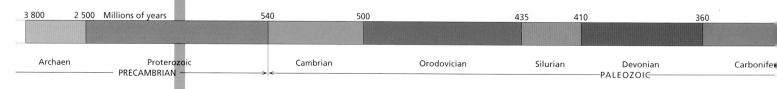

3 800	2 500 Millions of years		540	500		435	410		360
Archaen	Proterozoic		Cambrian	Orodovician		Silurian	Devonian		Carbonifer
	PRECAMBRIAN						PALEOZOIC		

Acidic Rocks to the North

Approaching the Beaujolais from the north, the first vineyards you notice are the ten top-ranking crus, all supported by the Hercynian basement: St.-Amour, Juliénas, Chénas, Moulin-à-Vent, Fleurie, Chiroubles, Morgon, Régnié, Côte-de-Brouilly and Brouilly. Granite formations are by far the most common and occur in the AOCs of St.-Amour, Juliénas, Chénas and Moulin-à-Vent; also in the upper part of the Fleurie cru, in the AOCs of Chiroubles and Régnié and in the Beaujolais Villages vineyards around Perréon, in the Vauxonne valley. Weathering of granite produces sandy, salmon-pink soils, known locally as "arène*" or "gore*" that can vary in thickness from a few inches to several feet. These poor, acidic, filtering soils are drought-sensitive but easy to work. The steepest slopes are found in the Fleurie and especially the Chiroubles appellations where the vineyards are planted on terraces separated by little walls of pink granite. The other acidic rocks that support the vineyards are Triassic sandstone in St.-Amour and schists of sedimentary volcanic origin in Juliénas. The personality of Beaujolais wines is defined by two formations in particular: the schists of the celebrated Côte de Py and the Carboniferous andesites* that form the Mont Brouilly. First, the schists. These weather to give blue-green soils rich in iron oxide and manganese that due to their color are known locally as "Morgon" ou *"terre pourrie"* (rotted earth). The Morgon appellation is one of the few AOCs to be named after the terroir (the Graves in the Bordelais is another). The Gamay in this iron-rich terrain gives a full-flavored, robust wine with a highly original bouquet of Kirsch plus ripe fruits with stones – so original indeed that the French invented the term "morgonner" to describe these unique organoleptic properties.

Let us now look at the andesites on Mont Brouilly. These are lavas that were metamorphosed by intrusions of neighboring granite. Côtes de Brouilly AOC wines (from the hill itself) are structured, tannic and long-lived, in sharp contrast to the supple, fleshy, delicate reds from the vineyards on other formations around the hill (granitic arènes, ancient alluvia, porphyry*, schists). The ancient bedrock in this northern part

The 1,400-foot-high chapel of La Madone, overlooking the pink granitic slopes of the upper part of Fleurie.

of the Beaujolais meets the Bresse plateau along a large fault. The many alluvial fans and accumulations of alluvia along this fault are due to weathering of the acid rock that breaks down into arènes (coarse sand). These are then washed away by rivers and deposited in vast alluvial fans where the bedrock meets the plain. They occur in the lower parts of Juliénas, Chénas, Fleurie, Brouilly and to the south and east of Morgon where the wines are rather less structured than those from the steepest slopes. Another important feature is the ancient alluvial terrace system formed by the Saône River that now flows alongside the vineyards.

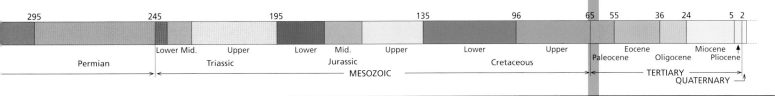

295		245		195			135		96		65	55	36	24	5 2

| Permian | Lower Mid. | Upper | Lower | Mid. | Upper | Lower | Upper | | Eocene | | Miocene |
| | | Triassic | | Jurassic | | | Cretaceous | | Paleocene | Oligocene | Pliocene |

MESOZOIC

TERTIARY

QUATERNARY

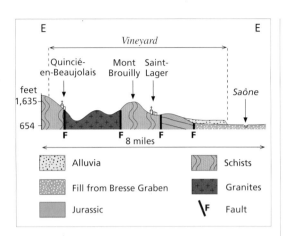

The windmill that symbolizes the appellation of Moulin-à-Vent is perched on a granitic hillock overlooking the vineyards.

On several levels and given over to the production of the Beaujolais Régionale wines, these terraces are often closely related to the alluvial fans and have partly reshuffled the sediments.

The edge of the Hercynian basement is easier to follow south of the Vauxonne valley, from St.-Etienne to St.-Julien and Denicé where it marks the lower limit of the Beaujolais-Villages viticultural area. This zone consists of a mixture of gneiss and granite around Arbuissonas, Le Perréon, cradle of the Beaujolais Nouveau, while Beaujeu, after which the region is named, is essentially schistose.

	Millions of years								
3 800	2 500		540	500		435	410		360

Archaen	Proterozoic	Cambrian	Orodovician	Silurian	Devonian	Carbonife
PRECAMBRIAN		PALEOZOIC				

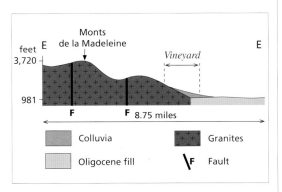

feet E
3,720

981

Monts
de la Madeleine

Vineyard

E

F   **F** 8.75 miles

	Colluvia		Granites
	Oligocene fill	\F	Fault

Golden Stones to the South

The eastern edge of the bedrock then rises toward the ridge of Pic Remont above Theizé. From the south of Mont Brouilly onward, the Mesozoic sedimentary formations reappear between the Hercynian basement and the Bresse Graben. Several ranges of stepped hills separated by secondary faults run from Pic Remont to Villefranche-sur-Saône. In the upper series, the Theizé region features significant outcrops of the Aalenian golden yellow limestone previously seen at the foot of the limestone ledge of the rock of Solutré. It was traditionally quarried as a building stone in the Bas Beajolais that came to be known as the *Pays des Pierres Dorées* ("Land of the Golden Stones").

The last chain toward the east separates the areas of Pommiers and Anse. Between the two extends the huge sloping ledge of Frontenas-Alix with its dense Upper Pliocene accumulations of detritus stripped from the hills of the Haut Beaujolais. This flatter wine-growing area produces the bulk of Beaujolais Nouveau, the fruity, quaffable wine that hits the stores the world over from November to December each year.

The Côte Roannaise and the Côtes du Forez

Several miles farther west there is a repeat performance of a geological phenomenon very like the formation of the Beaujolais hills, only simpler. The period of distension in the Oligocene saw the collapse of the Roanne and St.-Etienne basins in the Loire Valley. Then the Monts de la Madeleine and the Monts du Forez were uplifted, formed from granitic,

metamorphic rocks that created a front of rounded formations but with a rectilinear alignment from the north of the Côte Roannaise to the south of Montbrison. These formations then weathered, producing granitic arènes that were washed away in huge quantities and deposited at the foot of the slope on the edges of the Roanne and Forez trenches. Dividing the two (and the vineyards along them) are the Visean (Paleozoic) volcanic-sedimentary formations of the Neulize sill, slashed through by the spectacular Loire gorge and entirely free of vines. .

Mont Brouilly, home of full-bodied, tannic, Gamay wines.

What distinguishes the Côtes du Forez from the Côte Roannaise are the remnants of volcanic necks* dotted over the slopes and the plain. The Gamay is at home in both areas. Harvested later here than in the Beaujolais, it yields delicate, fruity, fresh wines with perhaps a touch more warmth and firmness on the Côte Roannaise.

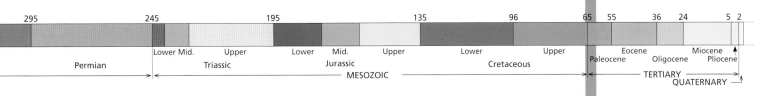

295		245		195			135		96		65	55		36	24		5 2

Permian	Lower	Mid.	Upper	Lower	Mid.	Upper	Lower	Upper	Paleocene	Eocene	Oligocene	Miocene
	Triassic			Jurassic			Cretaceous					Pliocene

MESOZOIC   TERTIARY
                            QUATERNARY

The Gamay Noir

Some grape varieties are forever associated with one particular viticultural area or wine. The Gamay Noir is one of them. Think of Beaujolais and you think of the Gamay Noir, even though it is cultivated much further afield than southern Burgundy.

The Gamay Noir accounts for a large proportion of French plantings but is especially famous in Burgundy thanks to the 14th-century French king, Philip the Bold. In 1395, being anxious to improve the quality of wine, the king issued a royal charter granting exclusive rights to the Pinot Noir and banishing the Gamay Noir, which he described as "a wicked and most disloyal plant". Winegrowers today know that the Gamay Noir can produce excellent or mediocre results, depending on the soil and crop yields.

The King of the Beaujolais Grapes

The Gamay expresses itself best in the two quite different styles of Beaujolais wines: on the one hand, Beaujolais Nouveau, the symbol of the region's commercial success; and on the other, Beaujolais-Villages and Crus Beaujolais. Beaujolais Nouveau is made by growers throughout the region although the heart of production is the *Pierres Dorées* zone in the southeast, an area of Triassic to Mid-Jurassic formations located west of Villefranche-sur-Saône.

The success of the Gamay depends on a special vinification technique based on the maceration of whole grapes and shorter periods of tank fermentation (three to seven days depending on the type of wine). This process brings out all the fruity intensity of the Gamay, producing crisp red wines with rich aromas of red berries, peaches, bananas, and flowers. Beaujolais Nouveau wines are soft and quaffable, like biting into fresh fruit.

The Beaujolais-Villages and Crus Beaujolais wines are made by the same vinfication technique, with longer periods of tank fermentation. They come from an entirely different terroir

dominated by granite and granitic arènes, plus more minor formations such as Triassic sandstone in St.-Amour, schists in the upper part of Juliénas and Morgon, and the andesites* in Brouilly. Gamay wines from the ten Beaujolais crus develop a different range of characteristics altogether. These are strong, full-bodied wines, combining all the natural fruit intensity of the grape with pronounced floral notes and rich, mouth-filling tannins. Some recall Burgundy wines from vineyards father north. Wines from steeper slopes where the vines are planted in the rock itself are for laying down; wines from the colluvial deposits at the foot of slopes tend to be lighter.

Gamay Plantings in the Rest of Burgundy

The Gamay is a very minor grape variety elsewhere in Burgundy, found only on the granite basement of the Morvan, beyond the Mesozoic limestone layer; also in the Couchey region and on outcrops of granite bedrock on the hills of the Mâcon area. There are also plantings in the limestone region of the Mâcon Rouge appellation, although the wines here never achieve the elegance of crus originating in granite terrain.

Gamay Plantings Outside Burgundy

Despite the Gamay Noir's mixed success outside Burgundy, there is one region at least, the Côte Roannaise, where it yields wines that rival the Beaujolais-Villages for fruitiness and quaffability. The terroir is similar (slopes of granitic arènes on the edges of the Loire Graben) but ripening is slower, producing wines with rather similar organoleptic properties to those of the neighboring Côte-du-Forez appellation, located slightly farther south on higher terrain.

Elsewhere in the Loire Basin, Gamay Noir plantings extend from the Auvergne (AOC Côtes-d'Auvergne) to the Atlantic Coast (VDQS Fiefs-Vendéens), with the exception of the Anjou-Saumur area where the Cabernet Franc takes over, and in the Sancerre area where the Gamay Noir is eclipsed by the Pinor Noir. In certain French wine-growing regions, the Gamay Noir is the only authorized grape for the production of *vins primeurs*. Touraine Primeur for instance is produced exclusively from the

The Côte Roannaise.
The Gamay from these
granitic slopes at the
foot of the Monts de
la Madeleine produces
fruity, fleshy wines not
unlike some of the
Beaujolais wines.

Gamay Noir. So too is Gaillac Primeur, made from vines planted in Tertiary gravel and clay deposits found along the edge of the Massif Central in the Cunac region, east of Albi.

In the Aveyron in southern France, the Gamay Noir grows alongside the Cabernet Sauvignon and the Cabernet Franc in the schists and granites of Entraigues-et-Fel and Estaing, and alongside the Syrah in the limestone *éboulis* (gravity slump material) of the Côte-de-Millau appellation. At the other end of France, it is blended with the Pinot Noir in the characteristic gray wine of the Côtes-de-Toul appellation, situated in the Oxfordian cuesta* to the east of the Paris Basin.

The Gamay Noir also thrives in mountainous regions. One of the best red wines of the Savoie region, for instance, is a fruity, easy-drinking wine called Chautagne Gamay from vineyards to the north of Lake Bourget. In Switzerland, we find plantings of Gamay Noir in the Valais Canton, especially in the acid soils of Fully and Martigny. The famous Swiss Dôle is produced from a mixture of Gamay Noir and Pinot Noir.

The Northern Rhône Valley

The narrow terroirs of the northern Rhône Valley extend along the Rhône River from Vienne to the south of Valence. Despite the southerly location, the climate remains continental with hot summers and cold winters. It may seem strange to learn that these terroirs are connected to those of Burgundy and Beaujolais. Geological evidence shows, however, that all three originated in the same process of collapse and uplift.

The hill of L'Hermitage above the town of Tain-L'Hermitage, viewed from Tournon (St.-Joseph appellation) on the opposite bank (cf. cross-section on the right). Visible in the background from left to right are the three sections that make up the hill. First there are the granitic slopes of Les Bessards, dedicated to red-wine production. In the center is the Quaternary terrace of the Méal area, planted with red and white grapes alike. Further down we find a mixture of loess and Quaternary pebbles in the vineyards of Rocoules and Les Murets, producing mainly white wine.

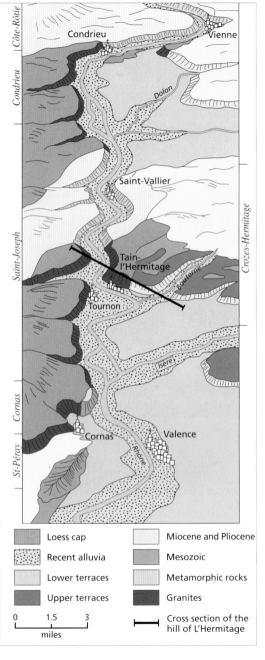

Loess cap		Miocene and Pliocene	
Recent alluvia		Mesozoic	
Lower terraces		Metamorphic rocks	
Upper terraces		Granites	

0   1.5   3
miles

Cross section of the hill of L'Hermitage

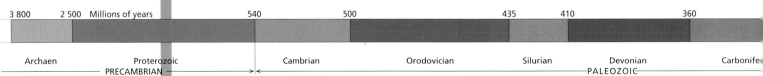

3 800	2 500	Millions of years	540	500		435	410		360

Archaen	Proterozoic	Cambrian	Orodovician	Silurian	Devonian	Carbonifer

PRECAMBRIAN

PALEOZOIC

The Lower Dauphiné, the area that now occupies the left-bank of the Rhône between Lyon and Valence, suffered subsidence* in the Oligocene just like the Bresse region. As it slumped, the depression became partly filled. Then the Mediterranean Sea launched an incursion into the Rhône Corridor and through to Switzerland, forming the pre-Alpine depression in the course of the Miocene. When the sea retreated, the Rhône was forced back to a line alongside the Massif Central by renewed tremors in the Alps. In the Messinian (Late Miocene), the Mediterranean Sea became cut off, dried up and the water level fell. The River Rhône dug itself a deeper bed along the edge of the ancient basement of the Massif Central and the sandy Miocene fill of the Lower Dauphiné, between Condrieu and St.-Vallier. It carved a deep gash in the granite basement in Vienne and to the north of Tain-L'Hermitage. Since then, the course of the river has remained largely unchanged thanks to two geological processes that helped it to maintain its position:

• The edge of the Massif Central was uplifted in the aftershock of the Alpine folding. This uplift was even more violent in the Beaujolais region, (where the Mont Pilat, overlooking the St.-Joseph appellation, reaches an altitude of 4,682 feet).

• Large quantities of material were washed away by the Isère River then deposited in the triangular area that extends from Lyon to Valence and Voiron. In the process, the Rhône River was forced back against the Massif Central and re-cut its banks to form a series of terraces (usually four). These have not survived on the steep right-hand bank along the Massif Central or in the narrow gorges at Vienne and Tain-L'Hermitage, but they are a characteristic feature of the landscape on the left bank and the Valence plain where they support vineyards (part of the Croze-Hermitage AOC) and orchards. Then in the Quaternary Period, violent winds stripped these terraces of their fine-grained silt that had accumulated as loess in areas that were then better sheltered.

The Côte-Rôtie, Côte Blonde, and Côte Brune

The vineyards of the northern Rhône are mainly located on the steep slopes on the border of the Massif Central. They are best illustrated by the vineyards of the Côte-Rôtie AOC ("roasted slope"). The uplift of the granite basement in the west did not occur along the central gutter of the Rhône Valley but in several compartments that are separated by faults running southwest-northeast. This is the direction corresponding to the major tectonic upheavals of the Hercynian Orogenesis that came back into play in the Tertiary. It explains why the Rhône suddenly switches course downstream of Vienne and starts to flow northeast-southwest along a fault, giving a near-perfect southeast exposure to the vineyards of the Côte-Rôtie. Metamorphic rocks form a steep scarp overlooking the Rhône River. They create the precipitous vineyard slopes of La Landonne, on the Côte Rozier, visible from the A7 Freeway just before the Rhône crossing south of Vienne.

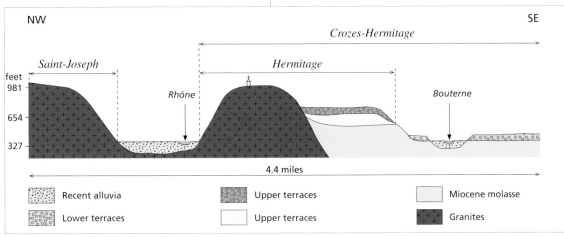

Section of the Hill of L'Hermitage.

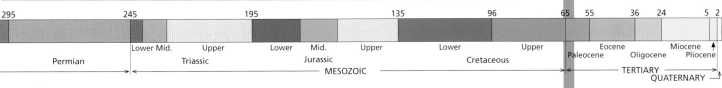

The steeply terraced
vineyards of the
Condrieu appellation
are planted with the
Viognier and extend
over a series of sites
including the Coteau
de Vernon (in the
foreground) and
the Coteau de Chéryl
(in the background).

The landscape is equally spectacular a short distance farther south, on the Côte Brune and the Côte Blonde overlooking the village of Ampuis. According to popular legend, the Lord of Ampuis named the two parts of the Côte after his two daughters, one blonde, the other brunette. Geology provides a less romantic but probably more authentic explanation. In fact, the two slopes are named after their different bedrock: the mica schist of the Côte Brune (Brown Slope) weathers to dark brown, iron-rich argillaceous soils; the gneiss of the Côte Blonde (Blonde Slope) weathers to lighter-colored, siliceous soils enriched with limestone loess* from the plateau. On these precipitous slopes, generations of patient vineyardists have had to build drystone walls called *cheys* to retain the soil.

The Syrah grape excels here, occasionally planted with small amounts of the white grape variety Viognier. Syrah wines from the Côte Blonde are soft and exceptionally fine and elegant, while those from the Côte Brune are robust, well structured and longer-lived. The red wines of the Côte-Rôtie rank among the best in the world.

Granites and Gneiss in Condrieu and Château-Grillet

Around Senons, the Rhône resumes its north-south course, while the granites and metamorphic rock to the south of Condrieu are exposed by the fault bordering the Côte Rôtie. This has significant consequences for the position of the vineyards. The eastern exposure seems less favorable here than on the Côte-Rôtie, and the slopes, although slightly less precipitous, remain steep. The streams that flow down from the Massif Central play a critical role. These short torrents, often flowing through faults and in the same direction, have left deep indentations along the edge of the plateau. Each valley mouth now provides a series of south-, southeast- and east-facing exposures that give the vines the sheltered growing conditions they require. One ideal vineyard location is the small, well-protected corrie of Château-Grillet, with its southern and southeastern aspects. The St.-Joseph appellation faces in the same direction on alternating granites and gneiss, criss-crossed with faults.

On the nose, aromas of peach, apricot and violets are overlaid by scents Condrieu and Château-Grillet were once the only vineyards to cultivate the Viognier, and the wines produced here from this little-known grape are quite unique. of honey in years of peak maturity. On the palate, there is a perfect balance of richness and roundness thanks to the grape's pronounced natural acidity. These wines, especially those from Château-Grillet, need to be properly aired in order to open up.

More Varied Formations around Valence

In St.-Vallier, the Rhône slips into another pass carved out of the granite bedrock. At the exit to this pass, the landscape opens out quite dramatically between Tournon and Tain-l'Hermitage onto a calmer embayment containing alluvial deposits from the Rhône. Only vestiges of the oldest, highest

3 800	2 500	Millions of years		540		500		435	410		360	

Archaen	Proterozoic	Cambrian	Orodovician	Silurian	Devonian	Carbonife
	PRECAMBRIAN				PALEOZOIC	

The Côte Blonde to the south of Ampuis features siliceous soils formed from gneiss, enriched by limestone loess from the plateau.

terraces survive here but they provide ideal locations for the vines. On the right bank are the vineyards of the celebrated St.-Joseph appellation, named after the hamlet of St.-Joseph, clinging to a fraction of stony terrace on the granite escarpment of the commune of Mauves.

On the opposite bank above Tain-L'Hermitage is the granite hill mass crowned by the famous chapel, a shield that saved a larger strip of ancient terraces from erosion in the most recent glacial periods. This strip extends all the way to the Bouterne Valley on either side of the freeway and is now occupied by the vineyards of Méal, Chante-Alouette, Les Murets and Rocoules. The Syrah gives of its best here, yielding powerful tannic wines with a nose of cooked fruits, spices and the forest floor. A Hermitage is rich and generous, with hints of spice, beeswax and toastiness, extending to animal notes with age. A Crozes-Hermitage is less concentrated, more supple and fruity, and intended for earlier drinking. Hermitage whites made from Marsanne and Roussanne have a floral bouquet. The Syrah grows in the granite terroir of L'Hermitage, while the white grape varieties, Marsanne and Roussanne, occupy the ancient terrace. To the southeast, the lowest (Würm glacial period) terrace at the confluence of the Rhône and the Isère

rivers is shared by the orchards and vineyards of the Crozes-Hermitage appellation. On the right bank, the St.-Joseph appellation stretches all the way to Châteaubourg. It is followed by the Cornas AOC, marked by a geological phenomenon typical of the vineyards of Burgundy and especially Beaujolais: a fault displacement that brings the granite basement into contact with the collapsed Mesozoic formations along the edge of the Lower Dauphiné.

Around Piélavigne at the northern end of the Cornas appellation for instance, the small Upper Jurassic (Kimmeridgian) limestone massif of Les Arlettes escaped erosion. It now provides an ideal shield against the north winds for an essentially granitic terroir producing powerful, full-bodied Syrah wines that contrast vividly with the softer wines of St.-Joseph.

These limestone formations disappear beneath St.-Péray only to resurface in the form of a spectacular escarpment dominated by the lofty ruined castle of Crussol. The Montagne de Crussol consists of Triassic formations overlaid by Upper Jurassic strata, and separates the present-day eastern course of the Rhône River from its former course in the west. This is where we find the deserted valley of the Toulaud, filled with Pliocene deposits at the foot of the St.-Péray vineyard.

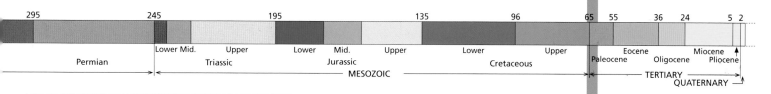

295		245		195			135		96	65	55	36	24	5	2
Permian		Lower Mid.	Upper	Lower	Mid.	Upper		Lower	Upper	Paleocene	Eocene	Oligocene	Miocene	Pliocene	
			Triassic		Jurassic			Cretaceous							
				MESOZOIC							TERTIARY				
													QUATERNARY		

The Syrah

The theory that grape varieties express themselves best at the northern limit of their cultivation is well illustrated by the Syrah that accounts for all red-grape plantings in the northern Rhône Valley. Most experts agree that the Syrah has been at home in this part of France for so long that it must have developed from native wild vines.

The northern Rhône Valley, with its Paleozoic and metamorphic rocks and dense, single cultivar plantings, has more in common with Burgundy than with Mediterranean vineyards. That said, there are no plantings of Syrah to be found anywhere north of Vienne, not even in the Beaujolais region, even though the terroirs there bear a certain similarity to those between Vienne and Valence (to the south).

The Northern Rhône Valley: Cradle of the Syrah

The topography and particular landscape here brings out a range of expressions in the Syrah. The view of the vineyards from the freeway at the exit to Vienne is quite spectacular. Leaving the industrial suburbs to the south of Lyon, the A7 skirts around Vienne on the steeply sloping right bank of the Rhône. Immediately before it rejoins the left bank, you can just see the precipitous slopes of the Côte Rôtie overlooking Verenay. Then the Rhône suddenly changes course and flows southwest along a large, southwest-northeast fault line.

Those first majestic, southeast-facing slopes of the Côte-Rôtie appellation, that extends from St.-Cyr-sur-le-Rhône to Semons on metamorphic formations, bring out the very best in the Syrah. The Côte Rôtie actually consists of a series of bluffs – the Côte Rozier, Côte Brune, and Côte Blonde – each separated by deep stream valleys that run down to the foothills of the Mont Pilat. There are as many terroirs here as there are styles of Syrah wine. Those of the two best-known Côtes, the Côte Brune and the Côte Blonde, are markedly different. Côte Blonde wines are soft, delicate, and exceptionally elegant while those of the Côte Brune are robust and solidly structured but take longer to develop. After Semons, the Rhône heads directly south once again. The Syrah gives way to the Viognier for a few miles, around Condrieu, before reasserting its rights in the St.-Joseph appellation alongside much smaller plantings of Marsanne and Roussanne. The curious St.-Joseph appellation originally consisted of six communes on the right bank of the Rhône

The Syrah flourishes in the Côte-Rôtie appellation, yielding soft, elegant wines on the Côte Blonde and robust, well-structured wines on the Côte Brune.

around Mauves, opposite the hill of L'Hermitage. Then in 1960, the appellation was extended to the north, toward Condrieu, and toward the plain and especially the plateau to the south, beyond Cornas and St.-Péray. Traditionally these more accessible areas had been reserved for the Gamay but as St.-Joseph's reputation grew, wine-growers also planted the Syrah. Then people in Lyon started to complain that St.-Joseph Syrah wines were no longer up to standard, so wine-growers swallowed hard and retreated to the steepest hillsides. Syrah wines from these precipitous slopes feature rich, delicate aromas, a very elegant palate and subtle tannins that emphasize the impression of roundness and meatiness.

A few miles farther south, the Syrah expresses itself quite differently in the neighboring terroir of the Cornas appellation. Here granitic arènes shielded from the north winds by the limestone bluff of Les Arlettes produce dark-colored wines with aromas of black fruits, blackberries and cherries. These are powerful, solidly built wines that take at least five to six years to mature. They are masculine wines, quite unlike the more feminine wines of St.-Joseph.

On the opposite bank of the Rhône, the terroirs of the Hermitage and Crozes-Hermitagere appellations are more varied; along the river, as in the St.-Joseph AOC, there are metamorphic rocks to the north and granites to the south. The Syrah predominates in the granitic, western part of L'Hermitage AOC but is less common on the series of terraces farther east that are formed of mixed alluvial deposits from the Rhône and the Isère. A significant part of the Croze-Hermitage AOC is located on the lowest terrace.

Syrah Hermitage wines are powerful and exceptionally elegant, with a wide range of aromatic nuances. Some are distinctly violet-scented, others are more peppery and some have a characteristic smokiness. Crozes-Hermitage wines are more supple and less concentrated, especially those from the lower terrace between the Isère and the Rhône where Syrah vines planted among the orchards are more productive than vines on the other slopes.

The Unstoppable Syrah Heads South

With the exception of Australia where it is now widespread, the Syrah has remained largely confined to its native Rhône Valley. Its expansion into vineyards farther afield in France, in the Vivarais, is very recent.

Syrah plantings increased when Mediterranean vineyards were granted AOC status in the mid-1980s. Its success here is best exemplified by the Faugères appellation, supported by Visean schists, and the Minervois-la-Livinière AOC where the Syrah has almost replaced the Grenache and the Carignan in the Lutetian limestone terrain of Agel.

In the New World, the Syrah has become a leading variety in Australia, where it is called the Shiraz. Generally, productivity is too high here to yield anything but mediocre wines, except in the Barossa Valley where growers have made great efforts to improve the quality of production. Some of these wines are now first-rate: powerful and spicy from the warmer zones of the Hunter and Barossa Valleys; more classical in style from the cooler, Coonawarra region (*cf.* p. 188).

While the majority of Australian Syrah wines tend to be fairly ordinary, those from the Barossa Valley are by contrast powerful, spicy and altogether exceptional.

Sedimentary Basins

Sedimentary basins are depressions in the Earth's surface between mountain ranges. Gradually, these basins become filled with an accumulation of two main types of organic matter. Basins in lower positions contain marine sediments of variable hardness (limestones or marls) deposited by marine incursions. Basins located on the edge of mountains or along faults (such as a graben) are filled with sediment of variable density created by the erosion of surrounding high land. This process is known as continental erosion. Often, these two types of sedimentation overlap.

Vines need sun and a favorable aspect. They also dislike excessive water and prefer filtering soils or sloping terrain with decent drainage. In principle therefore, the horizontal strata of sedimentary basins do not make the most suitable vineyard locations. However, a variety of geological processes can create very favorable sites.

For instance, where movement of the bedrock has modified the slope of the strata, the forces of erosion will have formed cuestas. These need a favorable aspect (as in the Paris and Limoux basins). The compact layers at the crest of the cuestas frequently disintegrate, covering the softer strata with eroded material (scree) that improves drainage and soil warming.

The geological features seen in a sedimentary basin may also result from the action of rivers that downcut into sediments, creating slopes and ridges along river banks that make ideal vineyard sites (as in the Loire Valley, the Libourne region, and Rioja). Sometimes where particular weather conditions apply—a warm, dry climate, for instance—vines may even be cultivated on almost flat terrain (as in Jerez) thanks to the moisture-retaining properties of the soil. This explains why vineyards are frequently located in sedimentary basins. Other significant geological formations are Quaternary terraces (in the center of a basin) and sometimes basement rocks and mountainous topography (around the basin). Both are the result of varying stresses caused by movements in the bedrock.

The Chablis vineyard

in the Paris Basin.

The Paris Basin

The Paris Basin is commonly regarded as the archetypal sedimentary basin, and has been studied by geologists from all over the world. It has a uniformly oval shape and opens onto the English Channel. To the west, it lies against the Massif Armoricain, to the south against the Massif Central and the Morvan, and to the east against the Vosges and the Ardennes.

The marine deposits of the Paris Basin accumulated over a period of nearly 200 million years, from the start of the Mesozoic Era (Triassic sandstone, followed by compact and soft Jurassic limestone strata) until the mid-Tertiary (Oligocene deposits of sand in Fontainebleau, in the center of the Basin). The marine incursion that deposited the chalks supporting the Champagne vineyards of the Marne was framed by just two brief periods of non-immersion at the beginning and end of the Upper Cretaceous.

Map showing vineyard locations in the concentric bands of the eastern Paris Basin.

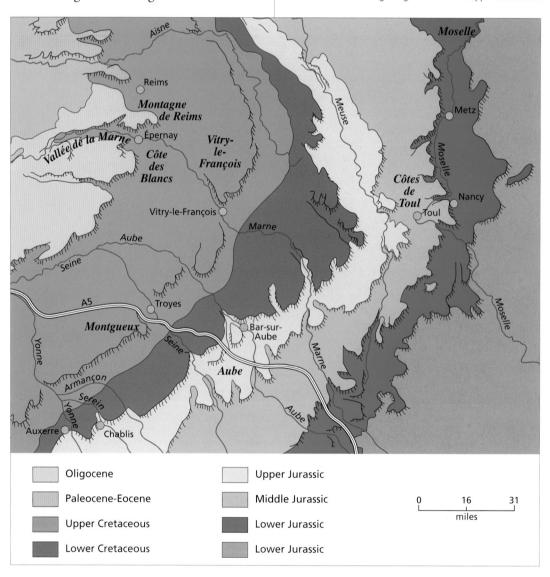

Legend:
- Oligocene
- Paleocene-Eocene
- Upper Cretaceous
- Lower Cretaceous
- Upper Jurassic
- Middle Jurassic
- Lower Jurassic
- Lower Jurassic

0   16   31 miles

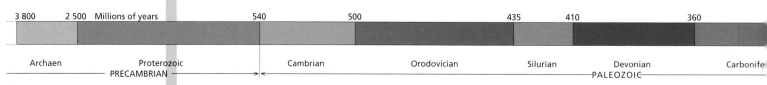

| 3 800 | 2 500 | Millions of years | 540 | 500 | 435 | 410 | 360 |

| Archaen | Proterozoic | Cambrian | Orodovician | Silurian | Devonian | Carbonife |

PRECAMBRIAN   PALEOZOIC

Sediments Deform the Basement

Throughout this period, the center of the Basin gradually sank deeper and deeper beneath the increasing weight of successive layers, to the point where the basement rock now lies more than 9,800 feet beneath the city of Paris. By the mid-Tertiary, the region had come to resemble a stack of nested dishes placed on a seriously collapsed basement.

If things had stayed that way, France might never have acquired any suitable vineyard locations. Fortunately, however, minor movements of the basement rock caused by the aftershocks of the Alpine folding, created niches that would later provide an ideal environment for first-quality vineyards.

The Basement Deforms the Sediments

The Morvan and the Vosges were uplifted from the end of the Miocene onward, as was the Bourgogne sill between them (in the Langres region). As the gradient of the sedimentary layers increased (in the range one to three per cent), so erosion chiseled away at the landscape to expose the hardest strata and produce cuestas* (scarp slopes) arranged in concentric bands. The rivers that now flowed toward the center of the Basin carved deep channels in these new landforms: the Marne in Epernay, the Aube in Bar-sur-Aube and the Seine in Bar-sur-Seine.

A Series of Cuestas

This cuesta topography is most evident from the freeway between Langres and Paris, marked by short, steep climbs and long, gradual descents. Roughly six miles out of Chaumont, just after crossing the Aujou River, the road climbs steeply onto compact Upper Oxfordian limestone where it twists and turns for another twelve miles. There is another steep climb beyond the Champignol rest area, onto Portlandian compact limestone, followed by a long gentle descent toward Troyes. At the foot of the Portlandian limestone lie the first Champagne vineyards of the Aube.

On a clear day, the next cuesta beyond Troyes is visible on the horizon. This is the chalk escarpment of Champagne that you drive down on leaving the freeway by the Troyes-Ouest exit,

passing the Champagne vineyard of Montgueux.

After crossing the Seine at Montereau, we see the Champagne vineyards of the Marne slightly farther east at the top of the Ile-de-France Ridge. From there the road gradually descends toward the outskirts of Paris.

This cuesta topography is especially visible from the eastern and southern parts of the Paris Basin. It is brutally interrupted at the level of the Loire Valley around Sancerre by a rift that first began to open up in the Paleozoic Era between the old blocks of the Massif Armoricain and the Vosges-Ardennes mountains. Toward the Lower Oligocene, this earlier crack resumed its downward movement, taking with it the overlying sedimentary layers. The Portlandian cuesta here is interrupted by a corridor several miles wide, through which flows the Loire in the section between Sancerre and Pouilly-sur-Loire, the corridor then continuing toward the Seine River Valley. Still farther west, the Bourgueil region and the southern section of the Saumur region were uplifted by the reactivation of ancient cracks in the Massif Armoricain that dives beneath the limestone layers of Anjou Blanc.

The Sancerrois appellation, surrounded by an amphitheater of Kimmeridgian marls (pictured here: the hamlet of Venoize, in the commune of Bué).

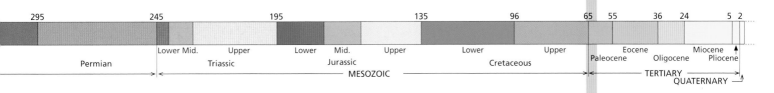

Champagne and the Lorraine Vineyards

The Champagne viticultural region is located barely 125 miles to the northeast of Paris, and extends into three departments: the Aisne and, in particular, the Marne and the Aube. There are also minor plantings in the departments of Seine-et-Marne and Haute-Marne. Extensive though it is (more than 90 miles at its longest point), the Champagne region is exclusively located on the favorably exposed borders of four cuestas in the Paris Basin.

Section of the Montagne de Reims.

Travelling eastward from Paris, the first cuesta is the ridge of the Ile-de-France. Around Epernay, the deeply incised trench of the Marne River valley separates the Montagne de Reims to the north from the Côte des Blancs in the south.

The Ile-de-France Ridge: The Montagne de Reims

The Montagne de Reims is a vast plateau capped by Sannoisian* (Lower Oligocene) millstones* that form the backbone of the cuesta and overlay all the Lower Tertiary formations (Paleocene and Eocene); and by the famous Campanian* chalk that is characteristic of the lower reaches of the slope. In the relatively compact Tertiary zone with its steeply wooded slopes, the chalk is mixed with material shed by erosion. The Pinot Noir and Chardonnay vineyards are mainly established on this chalky terrain, which may or may not be mixed with colluvia; there are also plantings in some of the Lower Tertiary formations. The Champagne produced here is first and foremost a blended wine, with aromas of red fruits and opulent body from the Pinot Noir, plus all the quintessential delicacy of the Chardonnay that adds crispness underscored by notes of citrus and toast.

To the north of the Vesle River that supplies water to the Petite Montagne, the vineyards occupy Lower Eocene sands. Here Lutetian* limestones form the backbone of the plateau.

The Côte des Blancs

South of Epernay, the cuesta continues in the Côte des Blancs. The Tertiary formations are much less thick here, leaving a chalky slope that is very steep in places and exclusively planted with the Chardonnay. Hence the name Côte des Blancs (literally "slope of the whites"). The delicate Chardonnay is at its finest and most elegant on these chalky slopes, bringing notes of butter, brioche and hazelnuts to young wines, white fruits (peaches, pears, quince) and honey to mature wines. Beneath it all is a peerless freshness and crispness due to the more noticeable acidity of the grapes from this area. Farther down the Marne River Valley, where the sedimentary strata tilt toward

3 800	2 500	Millions of years	540	500	435	410	360
Archaen	Proterozoic		Cambrian	Orodovician	Silurian	Devonian	Carbonife
	PRECAMBRIAN			PALEOZOIC			

the center of the Paris Basin, the chalk layer grows thinner toward the base of the slope, and then disappears below the Tertiary formations in Châtillon-sur-Marne, where the Pinot Noir takes over. Going eastward from Epernay, the chalky plains of cereal crops eventually lead to a cuesta of Turonian chalk. A number of recently replanted minor vineyards are situated along the edge, overlooking the villages of Vitry-le-François and Montgueux, to the west of Troyes.

The Côte des Bars

Farther east is the Aube district, the second key part of the Champagne viticultural area, where the vineyards enjoy sheltered locations in another cuesta: the Côte des Bars, home to Bar-le-Duc and especially, Bar-sur-Aube and Bar-sur-Seine. The dominant landscape features are formed from a hard limestone called the Portlandian (Upper Jurassic). This is the cap rock overlaying the Upper Kimmeridgian marls that support so many of the vineyards in the valleys of the Seine and the Aube, and the adjacent, east-west-facing valleys. The Chardonnay and the Pinot Noir tend to produce firmer, livelier wines here than in the Marne.

The slopes are relatively steep and whereas the Portlandian plateau is occupied by woods, the vines (Pinot Noir and Chardonnay) are planted in the Upper Kimmeridgian marls found on the steepest, upper part of the slope. The more calcareous, mid-Kimmeridgian at the base of the slope is given over to mixed farming. The Aube viticultural area does however

represent an exception from the geological point of view: at its southern tip, the small vineyard of Mussy-sur-Seine already lies on the edge of the next cuesta, a landform consisting of Upper Oxfordian limestones slashed through by the Seine Valley.

The Oxfordian Cuesta of the Lorraine

This Oxfordian cuesta may be regarded as a geological link between the Champagne viticultural area and the more modest vineyards of the Lorraine. Returning northward, you come to the Toul region where a small vineyard surrounded by Mirabelle plum trees occupies a sheltered spot in this cuesta: the Côtes-de-Toul appellation, famous for its *vins gris* dominated by the Gamay. These once quite acid wines are now better balanced thanks to the Pinot Noir, remaining lively but also very fruity and harmonious. In good years, the Pinot Noir yields red, often barrel-aged wines with notes of blackcurrant and vanilla. The white wines are dry and aromatic and made from the Auxerrois variety.

Our journey does not end there. There is still one more cuesta to cross, formed on the Bajocian limestone that predominates in Nancy and Metz, to the west. It is this cuesta that shelters the Moselle vineyard, planted in the Liassic marls along the river, a vineyard that continues into Luxembourg.

Diagram (cross-section):

NW — SE

feet
981
490

Woods

Vines

0.9 miles

- Portlandian
- Upper Kimmeridgian
- Lower Kimmeridgian
- Upper Oxfordian

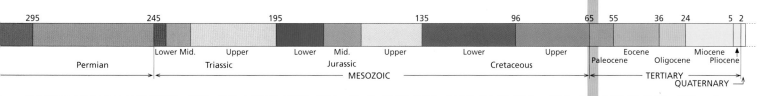

295		245		195		135		96		65	55		36	24	5 2

| Permian | Lower | Mid. | Upper | Lower | Mid. | Upper | Lower | Upper | | Eocene | | Miocene |
| | | Triassic | | | Jurassic | | Cretaceous | | Paleocene | | Oligocene | Pliocene |

MESOZOIC — TERTIARY
QUATERNARY

The Chablis
and Auxerre Viticultural Areas

With its vineyards extending into four bands of the Paris Basin, the Champagne viticultural area is an invitation to travel through geological time. The Chablis and Auxerre vineyards, on the other hand, share a single religion: the Kimmeridgian, topped by Portlandian cap rock. And in this particular context, use of the term 'religion' is deliberate.

The Côte des Bars represents a very regular westward continuation of the layers of Upper Kimmeridgian marls capped by hard Portlandian limestone that form the principal foundation of the Champagne vineyards. The commune of Les Riceys, famous for its rosé wines, is actually barely 18 miles from Tonnerre and some 25 miles from Chablis.

Chablis: A Geological Layer for Each Appellation

The slope of Chablis enjoys a generally southeastern aspect with a uniform composition in the following geological sequence, starting at the top:

• Lower Kimmeridgian *Calcaire de Tonnerre* (Tonnerre limestone): a chalky limestone only found in the southern part of the Chablis appellation.

• Lower Kimmeridgian *Calcaire à Astartes*: thick limestone (about 80 feet), so hard that it forms distinctive ledges in the relief.

• Mid- and Upper Kimmeridgian marls and limestone (260 feet), formed from alternating gray marls and limestone banks, some rich in the small, celebrated oyster fossil, *Exogyra virgula**. The slopes produced are tall and usually fairly gentle, although sometimes quite steep.

• Portlandian *Calcaires du Barrois*: very hard sublithographic* limestone deposits separated by banks that are richer in marls (163 feet).

The Chablis vineyard lies on Kimmeridgian marls overlaid with scree derived from hard Portlandian limestone. The limestone layer at the top of these slopes is wooded.

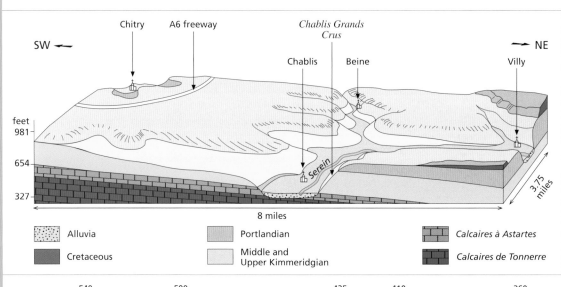

Chitry A6 freeway *Chablis Grands Crus*

SW ←   → NE

Chablis Beine   Villy

feet
981

654

327

Serein

3.75 miles

8 miles

| | Alluvia | | Portlandian | | Calcaires à Astartes |
| | Cretaceous | | Middle and Upper Kimmeridgian | | Calcaires de Tonnerre |

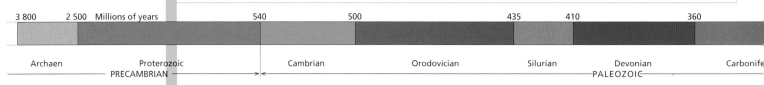

| 3 800 | 2 500 | Millions of years | 540 | 500 | 435 | 410 | 360 |

| Archaen | Proterozoic | Cambrian | Orodovician | Silurian | Devonian | Carbonife |
| | PRECAMBRIAN | | | PALEOZOIC | | |

74

The village of Irancy, nestling in a corrie of Kimmeridgian marls that opens to the southwest.

• Hauterivian limestone deposits (Cretaceous).
• Barremian marls, sands and clays (Cretaceous). Here the Portlandian limestone layers mark the crest of a *croupe** rather than a proper cuesta. These layers were uplifted by the northern continuation of the Hercynian bedrock of the Morvan, and tilt toward the center of the Paris Basin (toward the north-northwest at this point). What's more, like the River Armançon farther east (Tonnerre) and the Yonne farther west (Auxerre), the Serein River has made a deep indentation in the Côte des Bars, forming the slopes of Chablis into a vast V-shape that opens to the south-southeast.

The Four Appellations

The Chablis viticultural area is shared by twenty communes, divided into four appellation areas: Petit Chablis, Chablis, Chablis Premier Cru, and Chablis Grand Cru. The Chablis Grand Cru appellation is reserved for the best terroirs, exclusively located in the commune of Chablis. These are the seven *lieux-dits* (named vineyards) that overlook the village on the south-southwest aspect: Blanchots, Bougros, Les Clos, Grenouilles, Preuses, Valmur, and Vaudésir. They are all situated on brown, calcareous soils originating in Mid-, and Upper Kimmeridgian marls and argillaceous limestone, mixed with scree from the Portlandian ledge, as is often the case in this region.

The Chablis Premiers Crus lie along the Serein River, to either side of the Grands Crus on the right bank or facing them on the left bank. They are planted on the same formation as the Grands Crus, here interlaced with the Portlandian marly base. The remaining area is shared between Chablis and Petit Chablis.

A Single Grape Variety: The Chardonnay

The Chardonnay reigns supreme in the Chablis region where it develops lively, mineral qualities thanks to the combination of Kimmeridgian and Portlandian terrain, and a more northerly climate. The wines have a distinctive acidity that balances perfectly with the full-bodied texture and aromas of gun flint, quite unlike the honeyed or buttery nuances of the Côte d'Or wines. The Chablis appellation area, which is slightly more complex than Petit Chablis, brings out the fresh, floral (mint,

acacia, lime blossom) and fruity (lemon, grapefruit, green apples) qualities of the Burgundian grape. Chablis Premiers Crus and Grand Crus develop notes of dried fruits, almonds and St. George's mushrooms. The balance of richness and acidity is at its peak in these wines, aged in the wood or otherwise.

Throughout the appellation, right along the Côte des Bars, Chardonnay is blended with the Pinot Noir for the production of the Burgundy Appellation Régionale wines. Two sectors are particularly notable. The first is Irancy. The village itself, about nine miles south of Auxerre, nestles in a corrie that opens to the west. What is in effect a strip of horseshoe-shaped Portlandian limestone, cut off from the Côte-des-Bars by erosion, here protects the steep slopes of the subjacent Kimmeridgian. Irancy Pinot Noir, sometimes blended with the César, produces highly characteristic red wines with a rich, full-bodied palate and the backbone to withstand five to ten years' cellaring. With age, their bouquet of cherries, raspberries, black currants and blackberries takes on notes of spice, truffles, leather and the forest floor. In the neighboring commune of St.-Bris-le-Vineux, the Chardonnay partly gives way to the Sauvignon, especially on the northern sides, but only on the Kimmeridgian. The Sauvignon-de-St.-Bris appellation offers a foretaste of the vineyards of Central France.

Marly soils containing oyster fossils (Exogyra virgula) mixed with fragments of Portlandian limestone.

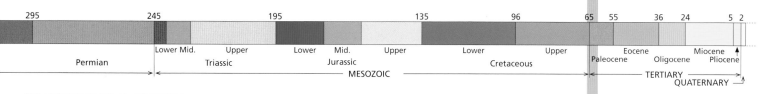

295		245		195		135		96		65	55		36	24		5	2
		Lower	Mid.	Upper	Lower	Mid.	Upper	Lower	Upper			Eocene		Miocene		
Permian			Triassic			Jurassic			Cretaceous		Paleocene		Oligocene		Pliocene	
						MESOZOIC							TERTIARY			
														QUATERNARY		

The Vineyards of Central France

Our trip along the Portlandian cuesta began with the Champagne viticultural area of the Aube, followed by the Chablis region, and now continues with the vineyards around Sancerre and Pouilly-sur-Loire. The vines are mainly established on Kimmeridgian marls and limestone that provide protection on two fronts: against erosion and against cold weather from the North.

In the Sancerre region, the uniformity of the cuesta is interrupted by a north-south trench some 6-7.5 miles wide that extends for nearly 190 miles toward Orléans and Rouen. The Loire River used to flow down this corridor into the English Channel via the Loing and Seine valleys. Its capture by the Atlantic Ocean and therefore its present-day course date from the end of the Tertiary Period.

The Sancerre Viticultural Area

This zone of collapse and the faults surrounding it have a major influence on the Sancerre region and, to a lesser extent, the vineyards of Pouilly-sur-Loire. The westernmost fault, with its virtually north-south orientation, runs along the western flank of the hill of Sancerre. It separates the following areas:

• To the west, the normal structure of the Paris Basin (that is, with the strata dipping toward the center of the Basin), composed here of the Portlandian cuesta overlaying Kimmeridgian marls and limestone, then Upper Oxfordian limestone.

• To the east, the collapsed block of Sancerre. Today this block stands out as a prominent landform due to a classical process known as relief inversion: the very hard, compact Eocene formations of flint scree and siliceous conglomerates that cap the block (at the top of the hill of Sancerre) proved more resistant to erosion than the Jurassic limestone and marly layers.

The village of Sancerre perches at the top of a hill capped by very hard Eocene siliceous conglomerates.

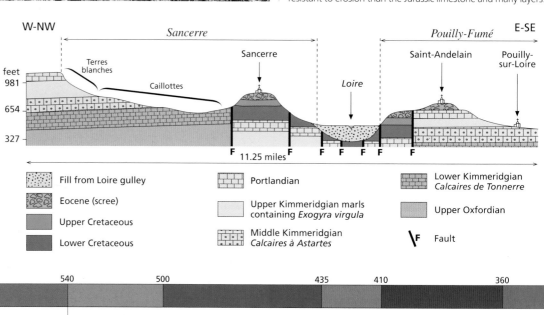

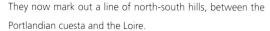

The Pouilly-Fumé appellation area on the borders of the Loire graben.

They now mark out a line of north-south hills, between the Portlandian cuesta and the Loire.

On the opposite side of the river, on the edge of the Loire, is the village of Tracy-sur-Loire, built on Eocene siliceous scree that has subsided to the height of the Portlandian in Pouilly-sur-Loire. The same Eocene scree also occurs on higher land, capping the butte of St.-Andelain.

The *Terres Blanches*

The Sancerre and Pouilly-Fumé AOCs have a profile quite unlike those of the Chablis and Auxerre appellations, even though they are located on the same band of the Paris Basin. This is a more complex geological situation where the nature of the wine depends on the nature of the formation.

Beneath the Portlandian caprock, which is more pronounced in Sancerre than in Pouilly, the uppermost slopes are formed from Kimmeridgian marls. These weather into soils known locally as *terres blanches* ("white earth"), that are rendzina type* or brown limestone, depending on the slope. The Sauvignon Blanc in this area produces full-bodied, firm, and solidly built wines with aromas that develop with age.

The *Caillottes*

Beneath these marls, lower layers of Kimmeridgian strata (Mid-Kimmeridgian – *calcaires à astartes** and Lower Kimmeridgian *calcaires de Tonnerre*), form an undulating landscape of very stony, rendzina type soils. Known locally as *caillottes or cris*, they are quicker to warm up than the *terres blanches*. Wines made from the Sauvignon Blanc are elegant, light and perfumed with a bouquet that develops and matures more quickly.

In the Eocene terrain and on the Cretaceous slopes of the butte of Sancerre, where the surface is covered with this siliceous scree, flinty soils yield firm, well-structured Sauvignon Blanc wines with that characteristic whiff of gunflint.

The Pouilly Viticultural Area

These types of terrain were once planted with the bulk of the Chasselas destined for the Pouilly-sur-Loire appellation. These days, the pattern of plantings has changed, with the Sauvignon Blanc now accounting for 95 per cent of the area under vine and producing the Pouilly-Fumé AOC. The soils are the same as in Sancerre and the Sauvignon Blanc wines are very similar: elegant and fragrant, with aromas of budding black currant in the cris; full-bodied and firm with nuances of narcissus in the marls; well-structured and rich with vegetal aromas or nuances of gunflint in flinty terrain.

The Menetou-Salon, Reuilly and Quincy Viticultural Areas Moving west, our trip through the Kimmeridgian leads to the Menetou-Salon appellation area, located in the continuation of the Sancerre AOC. Here the uniformity of the Kimmeridgian band is restored, and the vines grow almost exclusively on slopes of Kimmeridgian marls containing *Exogyra virgula**, those small oysters known locally as *œil de poule* ("hen's eye"). Here, the Sauvignon Blanc yields fresher, fruitier wines with aromas of oranges, lemons, quince and crisp apples.

Continuing westward, past Bourges and Cher, we come to the Reuilly and Quincy AOCs. The Portlandian cuesta has dropped away altogether here and no longer features in the landscape. In the Reuilly AOC, Kimmeridgian marls only sub-crop at a height of 490 feet (compared with 980 feet in Sancerre) beneath ancient Pliocene deposits of outwash.

In the Quincy AOC, the marls lie beneath Tertiary lacustrine limestones (Ludian and Stampian), themselves overlaid with ancient alluvia from the Cher known as the *sables roux de Castelnau* ('the red sands of Castelnau"). These thick alluvial deposits of sand and gravel some 16-19 feet deep yield Sauvignon white wines of exceptional finesse. This vineyard could be related to those on Quaternary terraces (*cf.* map on page 148).

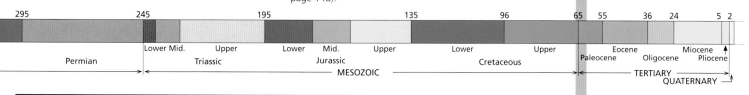

295		245		195		135		96		65	55		36	24		5	2
Permian		Lower	Mid.	Upper	Lower	Mid.	Upper	Lower	Upper		Paleocene	Eocene	Oligocene		Miocene	Pliocene
			Triassic			Jurassic			Cretaceous							

MESOZOIC
TERTIARY
QUATERNARY

The Sauvignon Blanc

In the Reuilly AOC, the Sauvignon Blanc is planted on much gentler slopes than in the Chablis or Sancerre appellations.

The Sauvignon Blanc covers such a vast area that it could almost be described as the universal grape variety. In France, the Sauvignon Blanc is mainly cultivated in the Loire Valley and southwest France, with lesser plantings in Provence and northeast France, in St-Bris-le-Vineux in the Yonne. It is equally widespread outside France and is now cultivated by most of the wine-producing countries of Western and Eastern Europe and North and South America, as well as South Africa, Australia, and New Zealand.

The taste of Sauvignon Blanc is immediately recognizable to any taster, no matter how inexperienced. So is that unmistakable nose of boxwood and budding black currant. Note, however, that these varietal aromas are less pronounced in wines from the Sauvignon Blanc's favorite terrain in Sancerre, Pouilly-Fumé, and Graves. Here, the grape develops a complex range of aromas that reflect the terroir and the skill of the winemaker.

A Versatile Grape

The Sauvignon Blanc is good for wines of any style, from dry to sweet or dessert wines. It is vinified alone in the single varietal wines of the Loire Valley, southwest France, and the Languedoc, and in those from that specific group of terroirs located on the Kimmeridgian band of the Paris Basin.

In St.-Bris-le-Vineux in the Yonne, the south-facing slopes are left to the Chardonnay while the Sauvignon Blanc occupies the northern slopes, formed from reddish argillaceous-stony colluvia mixed with Kimmeridgian marls. Here the ubiquitous grape produces lively wines with a mineral whiff of gun flint plus occasional tones of a Chablis.

The Sauvignon Blanc expresses itself best in the Sancerre and Pouilly-Fumé appellations on either side of the Loire River, where three kinds of terroir leave their particular mark on the grape:

• Kimmeridgian marls (or *terres blanches*), often steeply sloping, produce powerful, firm, full-bodied wines.

• Very stony soils (*caillottes** or *cris*) on Kimmeridgian limestone yield elegant wines with aromas of budding black currant and acacia.

• Siliceous, argillaceous-flinty soils bring out mineral and smoky aromas. Wines from these soils tend to have a fairly firm palate. In the Menetou-Salon and Reuilly AOCs farther west, Sauvignon plantings are confined to the Kimmeridgian, yielding elegant, very aromatic wines with a bouquet of citrus, blackcurrant, fern and acacia. Those from the sand and gravel terraces of Quincy are fine and crisp. Looking elsewhere in the Loire Valley, we find the Sauvignon Blanc predominating in the eastern part of Touraine around Blois, and in Cheverny on Loire terraces and Sologne sands.

The Sauvignon Blanc accounts for 15-20 per cent of plantings in the Monbazillac AOC, almost entirely planted on north-facing slopes.

A Legendary Trio

In southwest France, the Sauvignon Blanc accounts for almost the same planting acreage as in the Loire Valley (around 6,000 hectares). Its response to different teroirs however is less obvious since it is hardly ever vinified alone, being one of the legendary trio of Sémillon-Sauvignon Blanc-Muscadelle that reigns supreme in the white wine-producing regions of much of southwest France.

Sauvignon Blanc blended wines, whether dry, sweet or dessert, are among the best in the world. Due to its exuberant bouquet, the Sauvignon Blanc generally stands out in young dry wines, whereas the Sémillon only comes to the fore after three or four years aging in the bottle. This southern Sauvignon Blanc develops warmer, fruitier aromas than the Loire Valley vines. Take the white Graves, for instance: a perfect example of a wine that bears the mark of its terroir. Here, the gravelly slopes bring out aromas of white fruit, citrus and exotic fruit, followed by smoky, toasty characteristics in older Sauvignon Blanc wines. But the success of the Sémillon-Sauvignon Blanc partnership is not restricted to dry white wines. It is equally outstanding in the dessert wines, whether from the Sauternes-Barsac AOC on the right-bank of the Garonne River, or Monbazillac where the Sauvignon Blanc accounts for just 10-30 per cent of the blend. In gravel and limestone terroirs alike, the Sauvignon Blanc develops an aromatic intensity—particularly noticeable in young wines—together with a liveliness that promises good aging potential.

Antipodean Sauvignon Blanc: New Expressions in New Zealand The Sauvignon Blanc is now cultivated in countries all over the world. Outside France, it produces its best results in New Zealand, where it is now the number two wine grape variety. One outstanding example is Marlborough Sauvignon Blanc, from the central, largest viticultural area, yielding wines with a fruitiness and elegance unmatched elsewhere. This is the focus of Sauvignon Blanc cultivation, followed by Hawke's Bay and Canterbury.

The Sauvignon Blanc is the number two variety in the New Zealand archipelago, producing high-quality results.

The Touraine

The gently rolling landscape of the Touraine seems to radiate softness and tranquility from every feature. Even the geology seems to take it easy on this vast plateau, crossed by lazy rivers flowing indolently to the ocean, and dotted with timeless castles, large and small.

All the buildings in the Touraine, from grand châteaux to tiny loges (such as this one in the Jasnières AOC) were once built from soft, tuffeau limestone.

Because the ancient bedrock here was largely unaffected by movement, the strata are practically horizontal. The only features in an otherwise flat landscape are the valleys dug by the Loire and its tributaries. Following a prolonged period of non-immersion, the Upper Cretaceous Sea left behind the following legacy:

• A Cenomanian foundation of chalks, marls, sands and sandstone.

• A Turonian formation composed of Inoceramus* chalks (65-80 feet deep) and white and yellow tuffeau (respectively 130 feet and 65-98 feet deep), a soft limestone containing variable quantities of sandstone.

• A less extensive Senonian formation of clays, conglomerates and sands.

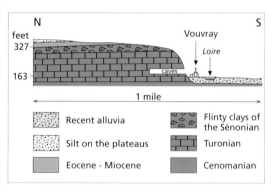

The Limestone Cliff of Vouvray.

The Touraine Trio

The Cenomanian, Turonian, and Senonian together form a trio that helps to explain the present-day topography of the Touraine. The landscape matured in the Tertiary Period, when the famous flinty clays were formed by surface weathering of the Senonian argillaceous-sandstone under tropical conditions. The grains of sand became cemented forming characteristic nodular clumps of all shapes and sizes that were used by the people on the borders of the Cher River to manufacture weapons, from Stone Age axe heads to the Meusnes gunflints used by Napoleonic troops.

The Eocene saw the formation of lacustrine limestone in Beauce, between Cher and Paris, followed by the Sologne

3 800	2 500	Millions of years	540	500	435	410	360
Archaen	Proterozoic		Cambrian	Orodovician	Silurian	Devonian	Carbonife
	PRECAMBRIAN					PALEOZOIC	

sands that the Loire and its tributaries had stripped from the uplifting granitic rocks of the Massif Central.

In the Quaternary, a pattern of consistent uplift spanned the entire region, causing the Loire and its tributaries to sink into Cretaceous sediments. The courses of these rivers were furthermore strewn with sandy and gravelly-sandy alluvia.

The Turonian

The words Turonian and Touraine are both derived from the name of a Celtic tribe called the Turones. Rarely has a geological formation left such a profound mark on an entire region. For centuries, the soft local limestone called *tuffeau* served as the local building stone, not only for the celebrated Loire châteaux but also for houses large and small, right down to those tiny *loges de vignes* ("wine lodges") to which the villages of the Touraine owe their characteristic neatness and brightness.

The Turonian can be easily identified in the landscape by the countless cave dwellings that were dug out of the yellow tuffeau along the borders of the Loir and Loire rivers, at the backs of the villages we see today. Indeed, the underground quarries that once supplied the stone now serve to cellar the wines that originate in the vineyards directly overhead.

Homogeneous Topography

We find the same landscape repeated right across the Touraine: vast, austere plateaus and narrow, pleasant valleys, some wider than others. In fact, the valleys of the Loir and the Loire and its tributaries follow much the same pattern.

The Cenomanian lies at the base of the series, occasionally outcropping at the base of the valley and frequently overlaid by ancient alluvia. Vines thrive in these warm, sandy-gravelly soils, such as we find south of the Montlouis appellation. On the sides of the valleys, the Cenomanian is crowned by the Turonian that usually forms a steep cliff riddled with caves that overlook the valley beneath. In some places however, where the tuffeau is less compact and the slope less steep, the vineyards are planted in stony clay-limestone soils known locally as "aubuis".

The classic flinty clays of the Senonian generally form the upper part of the valley sides. Most of the finer elements along the edge of the plateau have been stripped away by erosion, leaving very stony, flinty soils (perruches) favorable to viticulture. This kind of location is extremely common in the Touraine, especially in the Loir valley in the Jasnières, Coteaux-du-Loir and Coteaux-du-Vendômois AOCs. It also produces excellent results in the Loire Valley, in the AOCs of Touraine-Amboise, and Touraine-Mesland where the flinty clays are enriched by Miocene sands from the Massif Central. But the most outstanding example of this type of vineyard is the Vouvray AOC, a location that seems purpose-made for viticulture: plateau location, cellars dug out of the steep cliff face, and the Loire River at the foot of the cliff for shipping the wine to consumers.

To the east of the Touraine Appellation Régionale area, the Cheverny vineyard occupies a place apart. It extends over Beauce limestone, overlaid either by ancient alluvia along the Loire, or by Sologne sands. These latter formations support the Cour-Cheverny appellation that is exclusively planted to the fine, white and aromatic Romorantin grape.

The pattern of plantings in the Touraine has more to do with the weather than with geology and is broadly made up of the Cabernet Franc and the Chenin Blanc in the west, while the Gamay, Pinot Noir, and Sauvignon Blanc are more at ease in the east.

The terroirs of Cheverny and Cour-Cheverny.

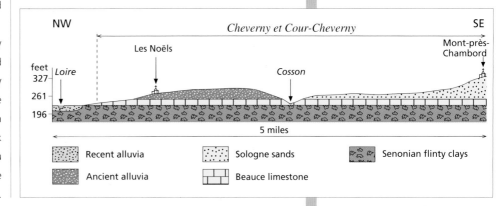

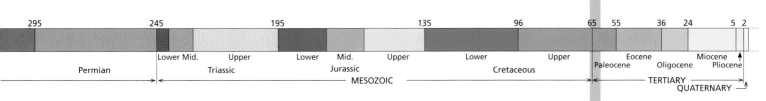

Western Touraine and the Anjou Blanc Region

Western Touraine comprises Bourgeuil and St.-Nicholas-de-Bourgeuil to the north of the Loire, and Chinon between the rivers Loire and Vienne. The Anjou Blanc region includes Saumur, to the south of the Loire and Vienne. These four appellation areas occupy the western tip of the Paris Basin, with the sedimentary layers of the Basin having been beveled away around Angers. At this point, the ancient Hercynian basement emerges to support the whole of the Nantes region and Brittany.

Movements of the bedrock also affected the distribution of the Mesozoic and Tertiary layers in this region, although not in the same way as at the other end of the Basin. In the Paleozoic Era, the Hercynian orogeny compartmentalized the southern part of the Massif Armoricain (Nantes region, Vendée) into a series of narrow parallel bands. Separating them are numerous faults, said to run in a south Armoricain direction (west-northwest, east-southeast), dipping beneath the Mesozoic formations of Anjou. These faults have been repeatedly reactivated since the Paleozoic Era, especially at the time of the Alpine Storm that had an impact even at this distance. Movements of the bedrock led on the one hand to the uplift of the southern part of the Bourgeuil appellation area (Chouzé-sur-Loire), and on the other hand to the uplift of the Montreuil-Bellay region, in the

Saumurois. In each case, the Jurassic limestone layers burst through the Cretaceous cover and outcropped at the surface.

Bourgueil and St.Nicolas-de-Bourgueil

These two appellations are entirely located on the slightly uplifted, domed Chouzé-sur-Loire block. Bordering it on either side are two faults, one to the south in Avoine, in the Chinon appellation area, the other to the east, in St.-Patrice, the premier commune of the Bourgueil appellation.

The Loire, which up to this point had dug a narrow valley (barely two miles wide) out of relatively hard Turonian limestone, now flowed through the more friable, argillaceous marls and sands uplifted along the edge of this fault. The proximity of the Massif Armoricain also had an impact on the sediments deposited in the Turonian and Senonian that were quite distinct from those in the Touraine. These were detrital formations, characteristic of the shallow seas and shores on the edge of the Massif Armoricain. Being much sandier, they proved far less resistant to erosion. These formations have had a dramatic impact on the landscape. The Loire Valley suddenly broadens in St.-Patrice and the valley sides, composed of Turonian and Senonian formations, fall away in a series of softer slopes that now support the AOC vineyards of Bourgueil and, to a lesser extent, those of St.-Nicolas-de-Bourgueil. The leading grape variety in the region, for the production of red wines, is the Cabernet Franc. Wines from different locations here are usually vinified and marketed

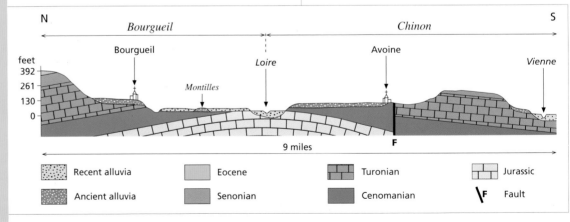

separately, allowing you to see how the Cabernet Franc responds to particular terroirs. On stony terraces, this late-ripening variety yields agreeable, elegant and aromatic wines (black currant, licorice) intended for relatively early drinking. Wines from tuffeau terroirs on Turonian slopes are fuller-bodied and more tannic and need four to five years to come into their own. The vines subsist frugally in Holocene alluvia, either on remnants of ancient alluvia (la Forêt, La Taille), or on delicate sand mounds known as *Montilles*.

The same formations also support the vineyards of the Chinon appellation, at the point where the Vienne valley, bedded into the Turonian, is bordered by towering cliffs several hundreds of feet high, as seen in the Loire Valley upstream of St. Patrice. Thus we find Turonian limestone ridges, source of the greatest Chinon wines of all, on the right bank of Crouzilles in Beaumont-en-Véron and on the left bank in Ligré and Roche-Clermault; ancient alluvia on two levels along the Vienne and part of the Loire at Savigny-en-Véron. The Cabernet Franc's responses to each type of terroir are the same as in the Bourgueil AOC.

The Saumur Region

The Saumur area covers practically the same formations as the Bourgueil AOC, except that it is divided into two by the Montreuil-Bellay fault, a huge scar in the landscape more than 90 miles long that extends into the Coteaux-de-Layon AOC.

The northern block of Saumur is raised up along this fault and tilted northward and toward the Loire. To the south, as the topsoil was gradually stripped away by erosion, the Jurassic limestone along the fault became exposed. Flinty clays eventually developed, and now support the vineyards of Brossay, Vaudelnay and Montreuil-Bellay. There are no vines at all in the somewhat sunken zone of Cenomanian sands and clays slightly farther north. Closer to the Loire however, from St.-Cyr-en-Bourg onward, there are extensive plantings on the hills around Champigny which are formed from Turonian and Senonian deposits. These formations support the Saumur-Champigny AOC, home of fleshy Cabernet Franc wines with good, solid structure thanks to dense but supple tannins, plus aromas of red fruits and spice. Unlike Bourgueil and Chinon, there are no ancient alluvia in this area; the Loire flows at the foot of the Turonian cliff, from Montsoreau to Saumur.

Soft and easily cut, tuffeau is ideal for building cellars. It is also porous and so helps to regulate ambient humidity.

The St.-Nicolas-de-Bourgueil AOC mainly extends over a stony Loire River terrace (pictured above, the Domaine de la Cotellerie).

The Saumur viticultural area.

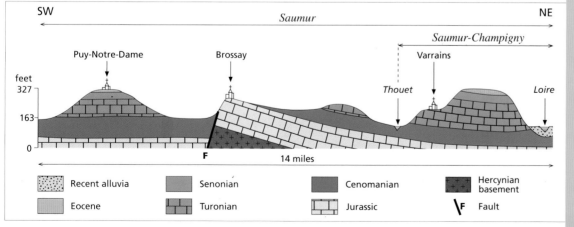

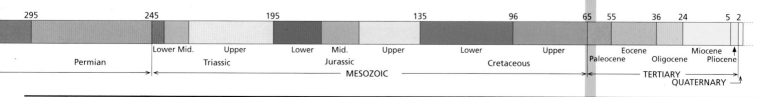

The Aquitaine Basin

The Aquitaine Basin is a huge depression located between the Massif Central, the Pyrenees, and the Atlantic Ocean. It is drained by the Garonne River and represents the second largest sedimentary basin in France. Within it are the regions of Aquitaine (meaning "water land") and Midi-Pyrénées, home of the Bordeaux, Bergerac, Cahors, and Jurançon AOCs.

The gray clays of the
Loupiac terroir.

The smoothly consistent Paris Basin emerged largely unruffled by the uplift along its rim plus a few cracks in the basement. The Aquitaine Basin by contrast has a violent history marked by three major phenomena.

• *The formation of the Atlantic Ocean*. At the beginning of the Mesozoic Era, the North American continent was joined to Europe. In the Jurassic, the two continents began to drift apart and the Atlantic Ocean was formed.

• *The opening of the Gulf of Gascony*. Until the Lower Cretaceous, the Gulf of Gascony was closed and the Cantabrian Mountains (northwest Spain) ran alongside the Massif Armoricain. Then the Iberian mini-plate went into rotation: it slid southeastward, then returned northward, colliding with present-day France.

• *The Pyrenean uplift*. Triggered by the movement of the Iberian plate, this shed erosional material that would cover the entire southern half of the basin.

What we are left with today is a basin that lacks any kind of symmetry.

An Asymmetric Basin

In the Triassic, the basin was divided into two along a line extending from Arcachon to Toulouse, with essentially shallow seas in the north and a deep sea in the south. Development of the northern part may be compared to that of the Paris Basin: relatively uniform sedimentation, minor movement of the bedrock, and uplift of the eastern border. In the southern part, the seabed underwent constant subsidence that persisted until a recent epoch. Today, the Hercynian bedrock rises to within 6,540 feet of the surface under Bordeaux but is nearly 23,000 feet down below Pau. At the start of the Jurassic, the waters of the Mesogean (future Mediterranean Sea) seeped in via the north of the Montagne Noire (Rodez region) and gradually invaded the basin. In the Mid-Jurassic, a barrier reef stretching from present-day Angoulême to Agen divided the Charentes area in the west, sinking under the weight of accumulated marls, from a shallow-water zone in the east. The massive limestones deposited in this eastern area would eventually form the *Causses*

3 800	2 500	Millions of years		540		500		435	410		360

Archaen	Proterozoic		Cambrian	Orodovician	Silurian	Devonian	Carbonife
	PRECAMBRIAN					PALEOZOIC	

The Château Cèdre vineyard in the Cahors AOC is established on the stony upper terraces of the Lot River, in Vire-sur-Lot.

(limestone plateaus) of Martel, Gramat, and Limogne. In the Lower Cretaceous, the seas retreated almost entirely except from two depressions (center of the Landes and the Adour basin). They returned with a vengeance in the Upper Cretaceous, depositing the calcareous chalks that support the Cognac vineyards. In the southern part of the basin, erosional material from the nascent Pyrenean range was shed into the north-Pyrenean trough.

Conflicting Influences

The Tertiary Period was deeply marked by the Pyrenean uplift* that led to the upthrust of the Massif Central. To the north of a line running from Bordeaux to Montauban, development was exclusively continental, with three principal types of deposit:

• *Sidérolithique**, produced by surface weathering of Jurassic and Cretaceous limestone under humid tropical conditions that filled the depressions with sands, gravels and iron minerals concentrated in a clay matrix.

• Gravelly clays washed down by rivers, from weathering of the Massif Central.

• Lacustrine limestone, especially at the foot of the Jurassic limestone plateaus.

To the south, development was typical of a region torn between two conflicting influences: on the one hand, erosional deposits stripped from the Massif Central and the Pyrenees,

washed down by rivers toward the center of the basin; and on the other, the Atlantic Ocean's relentless incursions, each one driven farther back toward the west.

The rivers flowed down both mountain ranges in torrents, rambling over an immense expanse that tilted toward the center of the basin. With them came an abundance of alluvia, sands and gravel that accumulated downstream, on the shoreline, in lagoon and delta zones that backed onto temporary lakes. Today, these deposits form a random collection of alternating layers of molasse and more or less hard limestone banks, essentially consisting of the following: Périgord sands from the Massif Central, that extend no farther than the Bergerac area (Eocene); Fronsadais molasse, lacustrine *Calcaire de Castillon* (Oligocene); *Molasse de l'Armagnac* (Late Miocene).

Not to be outdone, the Atlantic Ocean launched three major invasions. The first in the Eocene got as far as Blaye and Agen. The second in the Oligocene swamped the Gironde and the Landes. This is when the famous *Calcaire à Astéries* were deposited, one of the two major mainstays of the Bordeaux vineyards. The Miocene saw the formation of a huge lake (the origin of the gray limestone in the Agen area) stretching practically from Bordeaux to Agen. Then the sea launched a last, rather more timid incursion. At the end of the Tertiary, the region was blanketed in clays and gravels from the Pyrenees that spread over the land in the Pliocene as far as Bordeaux.

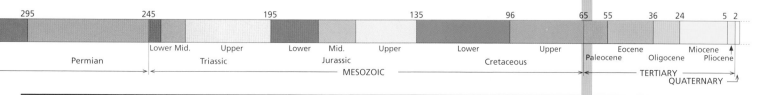

Cahors and Gaillac

The Clos de St.-Jean on the limestone plateau of Cahors (Clos de Gamot).

The best way to understand the vineyards of the Aquitaine Basin is to observe how each one became established, in chronological order, beginning with Cahors and Gaillac on the edge of the Massif Central. Following the dark days of the phylloxera epidemic, both these appellations in southwest France made a comeback in the second half of the twentieth century, thanks to the determined efforts of local winegrowers who focused on the concept of the terroir to save their vines.

The Cahors viticultural area covers some 4,200 hectares of land located in the meanders of the Lot River. The region is dedicated to the production of dark red wines, so deeply red that the English nicknamed them "black wines." Gaillac, on the Tarn River, covers a 2,687-hectare area and extends into five different terroirs. It is traditionally a white-wine area, producing dry, sweet and effervescent wines from particular local cultivars. Some Gaillac vineyards also make red and rosé wines.

Cahors: Limestone Massifs Carved by the Lot

The Cahors appellation area sits astride the Gramat and Limogne *Causses*, arid limestone plateaus mainly consisting of hard Kimmeridgian limestones that in this area are more than 1,200 feet thick. They outcrop extensively in the Limogne Causse to the south of the Lot Valley, where limestone beds alternating with marly limestones create a gently rolling landscape of shallow, very stony soils that are well suited to the Côt grape. To the south of the Cahors region, lacustrine

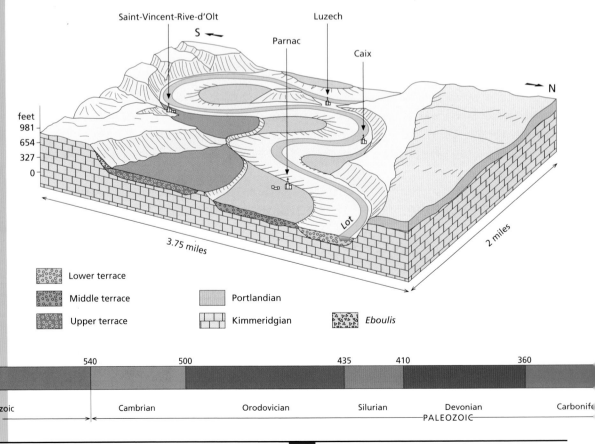

The Parnac and Luzech meanders in the Cahors AOC.

Lower terrace

Middle terrace

Upper terrace

Portlandian

Kimmeridgian

Eboulis

3 800	2 500	Millions of years		540	500		435	410		360
Archaen		Proterozoic		Cambrian	Orodovician		Silurian	Devonian		Carbonife
	PRECAMBRIAN						PALEOZOIC			

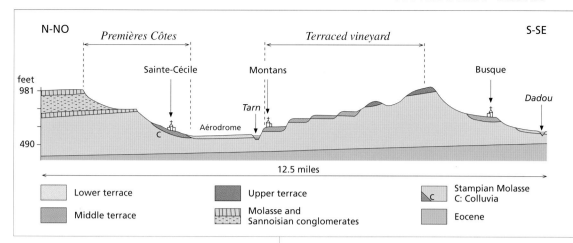

Vines at the foot of
Château de Mauriac,
planted in the Stampian
limestone and molasse
of the Premières Côtes,
in northern Gaillac.

limestone and patches of *éboulis* provide the only suitable vineyard sites along the ancient Kimmeridgian cliff, surrounded by Tertiary formations eaten away by erosion.

To the north of the Lot, the Kimmeridgian lies under some 230 feet of erosion-resistant, hard, gray Portlandian limestone, and there are hardly any vines at all here. They also avoid the area in the northwest, characterized by Upper Cretaceous limestone often overlaid by the *sidérolithique**. The Quaternary marked a major movement in the Earth's surface, which was to have a decisive impact on regional morphology. Following renewed uplift along the eastern part of the Massif Central, the Lot River dug a deep channel through the Kimmeridgian limestone, carving out its famous meander lobes (known locally as *cingles*). This is how the Cévennes Mountains got their steep, concave slopes. The inner loop was spread with alluvia that fanned out and formed staggered terraces separated by small heaps of talus. In the course of the Quaternary, the Lot deposited a series of three terraces, now occupied by vineyards. The highest, very stony and most-developed terrace is the most favorable to viticulture; the lowest, silty and sandy terrace is bordered by the Lot. The plateau edge is dotted with alluvial fans composed of limestone *éboulis* from small neighboring valleys.

Cahors is the domain of the Côt that accounts for more than 70 per cent of planting acreage and produces powerful, deeply colored, tannic wines. The finest vineyards are located on the upper terrace where the Côt performs best in soils mixed with variable quantities of *éboulis*. The plateau, where there were more vineyards prior to the phylloxera epidemic, also suits the Côt although when planted in limestone soils it is more sensitive to drought in summer and frost in winter.

Gaillac: An Open, Asymmetric Valley

As in Cahors, so also in Gaillac, where the Tarn has carved its channel through the Tertiary formations at the foot of the Massif Central in the Gulf of Albi. On either side are two Paleozoic massifs that were uplifted in the same period as the Pyrenees: La Grésigne in the north and the Montagne Noire in the south.

These are relatively friable formations consisting of gravelly clays in the Cunac area, and molasse overlaid by lacustrine limestone in the historic Cordes region. As a result, the valley formed by the Tarn (like the Vère and the Céron valleys farther north) is much more open than the Lot Valley and totally asymmetric. In effect, the steep south-facing right bank is cut by a series of ridges, limestone ledges, and hard conglomerates that form a semi-circle extending from Labastide-de-Livis to Rabastens. Behind these ridges, we find limestone and argillaceous-calcareous slopes on the limestone plateau and Stampian molasse of Cordes. This is the Gaillac Premières Côtes region, an area of white wine production, particularly suited to the local grape varieties, Len de l'El, Mauzac, and Ondenc. The north-facing left bank supports a series of stepped terraces separated by heaps of talus, varying in height from 49-65 feet, which formed by deposits of Quaternary alluvia from the Tarn. These discontinuous terraces, ribbed by small erosional valleys at right angles to the main valley, are the domain of the red grape varieties, especially the Fer-Servadou. To the south of the area, the Dadou River, like the other tributaries in the region, also dug an asymmetric valley. This asymmetry developed because the east-facing slopes that caught the sun warmed up more quickly in the glacial periods and were more susceptible to erosion than the north-facing slopes.

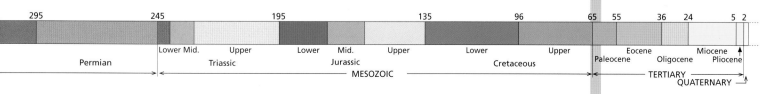

The Bergerac Viticultural Area

T he Dordogne vineyards offer a foretaste of the Libourne appellations. The Dordogne River, a tributary of the Garonne, does in fact flow past a succession of viticultural areas on its way to the Gironde: Bergerac, Pécharmant, Rosette, Monbazillac, Saussignac and Montravel. The Bergerac region, in the southwestern part of the Dordogne, formed the top of the department before the boundaries were redrawn, its wines traveling downriver to Libourne and Bordeaux.

The Bergerac appellation area is characteristic of a region exposed to the conflicting influences of marine deposits on the one hand and continental deposits on the other. Witness the series of deposits:

• Marine sediments (Eocene marls and clays, Oligocene *Calcaire à Astéries*);

• Continental deposits (Périgord sands, Fronsadais molasse, Agenais Molasse);

• Lacustrine limestone (*Calcaire de Castillon, Calcaire de l'Agenais*).

On the Right Bank of the Dordogne

To the north of the waterway lie two quite distinct zones. The first extends to the north and east of Bergerac, on hummocky terrain chiefly consisting of Périgord sands but with outcrops of Upper Cretaceous limestone in some valley bottoms and on the cliff overlooking Creysse. The argillaceous, compact soils here are mostly unsuitable for viticulture except in small amphitheaters and on the *croupes* along the edge of the Dordogne Valley. It is here that we find the Pécharmant AOC, an area dedicated to the production of red wines from soils that contain clay and iron minerals leached from the surface layer. This leaves a water-resistant clay and iron layer called *tran* in French. It is rich in ferruginous concretions known

The croupes of Saussignac overlook the Dordogne River.

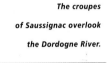

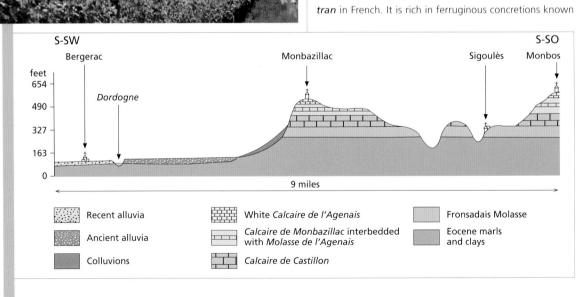

Cross-section, Monbazillac-Monbos, in the Monbazillac AOC.

*Monbazillac castle,
perched on a hill
of lacustrine
Calcaire de l'Agenais.*

locally as "goats' droppings" that are said to give a Pécharmant its taste of the terroir. The Merlot and the Cabernet Sauvignon are planted in sands and gravels mixed with Cretaceous soils at the base of the slopes, and yield robust wines with prominent tannins.

The second zone on the right bank is occupied by the Montravel sub-regional appellation area that extends from the Fleix river meander to the departmental border with the Gironde. The Périgord sands never made it this far, whereas the Stampian Sea (Oligocene) invaded the entire region from the west, from Montravel to Fleix.

The hill overlooking the Dordogne Valley is formed from Fronsadais molasse capped by lacustrine *Calcaire de Castillon* that stands out in the landscape as a wooded line on a whitish ledge of chalky limestone. The limestone contains *Calcaire à Astéries* deposited by the Stampian Sea. The layer is very thin around Fleix but grows progressively thicker toward the west as you approach the Gironde department (Côtes-de-Castillon AOC followed by the St.-Émilion area). The limestone on the plateau is overlaid with *Molasse d'Agenais* that forms buttes, some of them capped by a strip of lacustrine *Calcaire de l'Agenais* (Moulins de Ponchapt).

These warm, stony soils of *Molasse de l'Agenais* and *Calcaire à Astéries* make the best vineyard sites. The dry wines originate from the foot of the slope in the Dordogne Valley while the sweet wines come from the plateau edge.

On the Left Bank of the Dordogne

The left bank of the Dordogne, to the south of the river, looks more uniform, even though the topography is actually very rugged due to deeply eroded Tertiary formations. The formations are the same here as in the Montravel AOC but without the *Calcaire à Astéries* since the Stampian Sea did not make it this far. As a result, the *Molasse de l'Agenais* rests directly on *Calcaire de Castillon*. This is the home of two famous appellations renowned for their white dessert wines: Saussignac in the the west, and especially Monbazillac, in the east, on the edge of the Dordogne Valley, bathed in early morning mists followed by autumnal sunshine.

The vineyards in the Monbazillac AOC begin on a Dordogne Quaternary terrace, in St.-Laurent-des-Vignes, then ascend by degrees to alternating layers of limestone and molasse. A small plateau of lacustrine *Calcaire de l'Agenais* forms the caps of the buttes of Monbazillac, Thénac and Monbos. The Saussignac vineyards are mainly established on gravity slump material of limestone Grèzes* and boulbènes* from the plateau. In both types of vineyard, the Sauvignon Blanc, Sémillon, and Muscadelle occupy mainly north-facing sites—an exceptional location for these varieties.

The Saussignac and Monbazillac hinterland is the domain of Bergerac red wines. The appellation vineyards extend over the same formations as those described above, plus Quaternary alluvia along the Dordogne River.

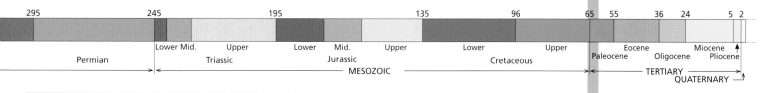

The Libourne
and Entre-deux-Mers Areas

Garonne and the Dordogne, is a vast hilly expanse planted with 23,000 hectares of vines.

The Libourne region (or Libournais as it is known in France) with the busy commercial town of Libourne at its center, occupies a particular place in the Bordeaux hierarchy. The vineyards are predominantly planted to the Merlot grape and concentrated around a handful of famous communes: Fronsac, Pomerol, and St.-Émilion. By contrast, the Entre-Deux-Mers region, at the confluence of the

The long, straight, departmental roads joining Bergerac to Libourne provide an excellent demonstration of the transition between Montravel and the St.-Émilion area. They show clearly the extremely straight geological layers that make up the region. The scarp of Fronsadais molasse, capped by a limestone ledge, runs continuously from St.-Foy-la-Grande to the departmental boundary. It is then broken up by broad valleys around the Côtes de Castillon AOC, before running straight and smooth once again in St.-Émilion. The scattered vineyard slopes of the Dordogne make way for denser plantings around Castillon-la-Bataille leading to vine monoculture in St.-Émilion.

The Côtes-de-Castillon AOC

This is the first appellation you come across as you enter the Gironde. It produces red wines from vines planted on the same formations as in Montravel: Fronsadais molasse capped by lacustrine *Calcaire de Castillon* and *Calcaire à astéries**. In

The village of St.-Émilion is built on Calcaire à Astéries.

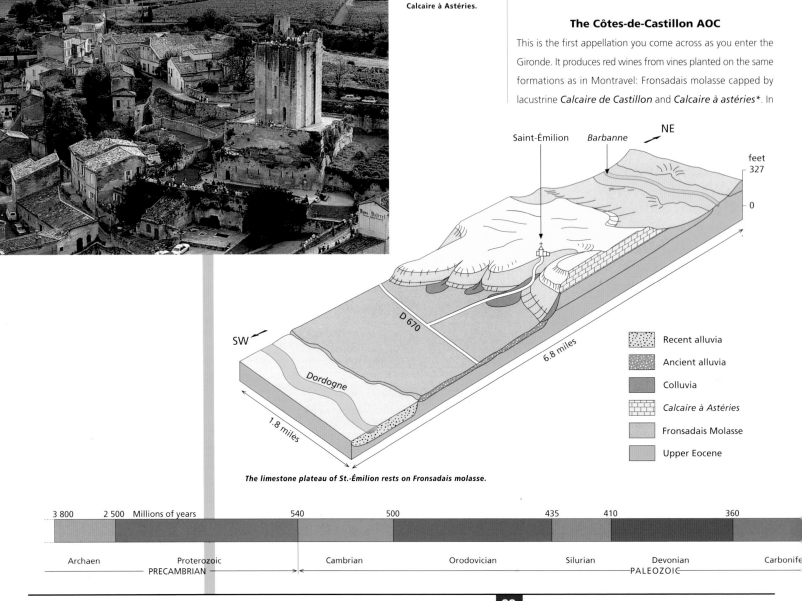

The limestone plateau of St.-Émilion rests on Fronsadais molasse.

The vineyards of Château Ausone, on the edge of the plateau formed from Calcaire à Astéries, are planted in Fronsadais molasse.

this area however, the plateau zones are interrupted by broad valleys, at the bottom of which lie the villages and, in many cases, crops other than grapes. The wines are produced from a blend of three grape varieties, the Merlot, Cabernet Sauvignon, and Cabernet Franc. Vines planted in the lower section yield supple, warm and delicate wines, whereas those on the plateau produce more robust wines with confident tannins.

At the border between the Côtes-de-Castillon and St.-Émilion appellations, the lacustrine *Calcaire de Castillon* disappears. It is replaced by clays, marking the limit of that vast Oligocene lake that once extended as far as Monbazillac.

The Plateau and Slopes of St.-Émilion

The legendary slopes of St.-Émilion are a perfect illustration of regional morphology. The Stampian Sea left behind it a huge plateau of hard limestone that now extends across most of the department of the Gironde. Around St.-Émilion, this plateau was deeply incised by the rivers Dordogne in the south, Barbanne in the north, and Isle in the west. All that remains of it today is a small, battered ledge of limestone 196-294 feet high, dissected by erosion around the village. Perching proudly on its southern edge is the village of St.-Émilion.

The villagers did not have far to go for their building materials: the local *Calcaire à Astéries*, soft enough to cut easily but tough enough to withstand the ravages of time, has been used in most of the buildings in the Gironde. It is what gives St.- Émilion its characteristically warm yet elegant appearance. These days, the galleries from which the stone was once extracted are used for wine storage.

The soils of the plateau – reddish scree and limestone fragments mixed with decalcified clays – are derived from weathered limestone. They are one of the mainstays of the

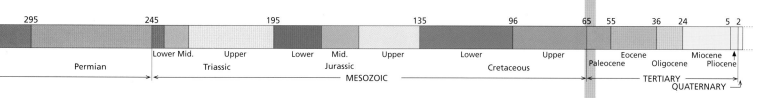

The Château Figeac vineyard occupies the stony upper terrace on the borders of the Pomerol AOC.

vineyard, although not the only one. The flanks are covered with scree that slipped down from the plateau. There, mixed with Fronsadais molasse and sometimes Eolian sands, it now forms an excellent, well-drained growing medium for the vines, especially in the vineyards on the steepest part of the scarp, south of the village, overlooking the Dordogne River. The vineyards here receive a constantly supply of water thanks to fractured (porous) *Calcaire à Astéries* overlaying impermeable clays in the upper portion of the Fronsadais molasse. Rain water percolates into the *Calcaire* then rises to the surface, creating springs where the porous *Calcaire* meets the impermeable molasse. This is the origin of the two springs in St.-Émilion, La Médaille and the Puits des Jacobins.

Given this combination of steep slopes, south-facing aspects and perfectly balanced soils, most of the St.-Émilion Grands Crus Classés are, as you might expect, located at the southern edge of the plateau. Not exclusively, however. The appellation also extends into two further sectors that depart from the usual pattern of plateau and molassic slopes.

• The Dordogne Plain, with its very wide flats of river-deposited Quaternary gravels and sands. The Libourne section of the plain was formerly the Sables-St.-Émilion AOC (one of the few appellations to be named after its terroir) before it was absorbed within the St.-Émilion appellation in 1977. The wines from the plain are pleasingly supple and fruity.

• The stony terrace at the edge of the Pomerol AOC. The highest, stoniest Pomerol terrace extends well beyond the administrative limits of the commune, onto the western border of St.-Émilion. This is where we find Châteaux Figeac and Cheval Blanc. In this particular terroir, the Cabernet grape varieties take over from the Merlot, which predominates in the rest of St.-Émilion, yielding long-lived, powerful, and well-built wines. In fact, this and the southern edge of the plateau are the only terroirs perfectly suited to the Cabernet Sauvignon.

Setting aside these two sectors, the surrounding appellations follow exactly the same pattern as the St.-Émilion AOC. Its "satellite" communes, to the north of the Barbanne River— Lussac, Montagne, St.-Georges, and Puisseguin – occupy identical sites on a plateau of Calcaire à Astéries surrounded by valleys carved out of the subjacent molasse and covered with éboulis. The big difference is that they do not have St.-Émilion's magnificent location on the edge of a scarp. Furthermore, the landscape is no longer dominated by viticulture.

3 800	2 500	Millions of years		540		500		435	410		360
		Archaen	Proterozoic		Cambrian		Orodovician		Silurian	Devonian	Carbonife
		PRECAMBRIAN							PALEOZOIC		

Fronsac and Canon-Fronsac

The pattern of "*Calcaire à Astéries*-Fronsadais molasse" also applies to the Fronsac AOC. Here however the limestone plateau at the top of the scarp is much more jagged and eroded due to its location at the confluence of the rivers Isle and Dordogne. But despite the landscape of steep-sided hillocks, the soil sequence is the same as in St.-Émilion. The Canon-Fronsac AOC, on the southern side of the Fronsac area, does in fact recall the spectacular location of the St.-Émilion Grands Crus perched on the southern edge of the plateau.

Entre-Deux-Mers and the Premières Côtes

The Entre-Deux-Mers vineyards (dry white wines) are established on Oligocene, Miocene and Pliocene strata, supported by a framework of *Calcaire à Astéries*. This is the main formation even though it gradually slopes toward the southwest, dropping from the top of the scarp on the left bank of the Dordogne to the foot of the St.-Croix-du-Mont cliff, along the Garonne River. This cliff represents all the formations found in the Entre-Deux-Mer area, from the Oligocene *Calcaire à Astéries* and gray clays at the base, to the Pliocene gravelly clays at the top. In between lie Miocene sandstone deposits overlain by the oyster bed (shallow marine level) that forms the bedrock of the castle and church of St.-Croix-du-Mont, and also the face of the scarp.

Together, the Dordogne and Gironde rivers have chiseled away at the stack of strata right down to the Calcaire à Astéries, leaving a deeply jagged outline. Most of the hummocky hills are formed of gray clays that tend to be unsuitable for viticulture. The Miocene Sea mainly invaded the southern area and barely went beyond the Sauveterre-de-Guyenne to Créon road. It left behind a landscape strewn with small buttes, overlain with Pliocene gravelly clays such as we find in Gornac. These are well-suited to the production of red wines that are gradually overtaking the whites.

The Côtes-de-Duras AOC with its deeply incised and eroded plateau, especially scarred by the Dourdèze River, provides a transition between the Entre-Deux-Mers and Bergerac regions. The formations are the same as in Monbazillac but with the addition of *Calcaire à Astéries* that is usually clearly visible along the plateau edge. The best vineyards, around the town of Duras, lie on *Calcaire de Castillon*. The Côtes-du-Marmandais AOC is essentially a continuation of the same terroir, although the appellation itself overflows onto the left bank of the Garonne. There, the sands of Les Landes skirt the upper river terraces around the town of Cocumont.

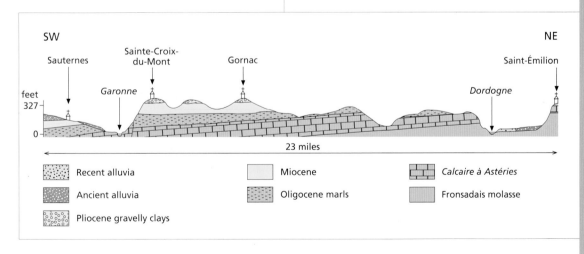

Cross section,
Entre-Deux-Mers.

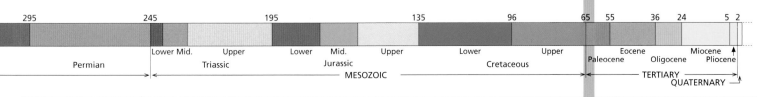

The Merlot

The Merlot, the Cabernet Franc and the Cabernet Sauvignon form a well-known trio of Bordeaux grape varieties. Irrespective of the terroir or the country of cultivation, all attempts to imitate the quality of the great red clarets of the Gironde have invariably been based on a blend of the Merlot and Cabernet varieties. Despite their close association however, outside the Gironde each of these three grape varieties has fared quite differently.

The Merlot and the Cabernet Sauvignon are each estimated to account for some 150,000 hectares of planting acreage worldwide. Their respective distribution patterns are different however, the Merlot being centered on the Gironde and France in general, while the Cabernet Sauvignon is more cosmopolitan. The figures speak for themselves: 45,000 hectares of Merlot plantings in the Gironde, representing one third of Merlot acreage worldwide, compared with 25,000 hectares of Cabernet Sauvignon plantings, or one sixth of its worldwide distribution. What's more, barely half of all Merlot plantings

Above:
The Merlot accounts
for 65 per cent of
plantings in the
Château l'Evangile
vineyard in Pomerol,
where it grows in
impermeable clays and
argillaceous gravels.

Left:
The rolling vineyard
slopes of Fronsac meet
up with the Dordogne
River in places.

are outside France, compared with nearly three-quarters of registered Cabernet Sauvignon vines.

King of the Gironde

The Merlot is the most widely planted red grape variety in the Gironde and continues to expand its dominance in the red wine category. For many years, the Cabernet Sauvignon was actually the main authorized grape variety for all new sites approved by the Institut National des Appellations d'Origine and the Ministry of Agriculture. It started to lose ground in the mid-1980s and has been in decline ever since. This is partly due to its disappointing performance when planted in unsuitable locations, and partly to certain inherent drawbacks compared to the Merlot. The Cabernet Sauvignon is more prone to Eutypa die-back, for instance, an increasingly prevalent and persistent fungal disease that reduces the sugar content in the musts. The Merlot, on the other hand, is less tannic and lighter bodied and therefore suitable for earlier drinking, a significant advantage given the storage overhead entailed in aging wines.

Château Bellevue
(St.-Émilion Grand Cru)
is dominated by the
Merlot (90 per cent
of planting acreage),
growing in argillaceous-
limestone soils on
a bedrock of Calcaire
à Astéries.

Merlot plantings have considerably increased in recent years, even in the Médoc, where in principle the Cabernet Sauvignon is supposed to do best on the Quaternary gravel terrace mounds. But even here there are signs of a Merlot take-over, the Cabernet Sauvignon and the Merlot respectively accounting for 48 and 47 per cent of plantings in the Médoc, and 52 and 41 per cent in the Haut-Médoc. We might see a change in the characteristics of Médoc wines.

Deep, Cold Soils

Apart from the Médoc area, the Merlot and Cabernet Sauvignon generally occupy different locations in the Gironde, for reasons to do with the terroir and the weather. In fact, each variety requires quite distinct growing conditions. The well drained, warm, stony gravels that are such a favorite with the Cabernet Sauvignon do not always suit the Merlot, which prefers deeper and especially colder soils. Often unbalanced in warmer soils, the Merlot positively shines in cool argillaceous to argillaceous limestone terrain, where it produces results to rival the very best Cabernet Sauvignons.

One such terroir spans the St.-Émilion and Pomerol AOCs, on the banks of the Dordogne River. The regional geology here is fairly straightforward: Fronsadais molasse foundation with somewhat heavy argillaceous-limestone soils. The molasse is overlain – particularly around the village of St.-Émilion – by the famous *Calcaire à Astéries*, which forms the plateau summit and gives rise to argillaceous-limestone soils, usually rich in gravel. In Pomerol, the Fronsadais molasse is overlain with scree aprons from the alluvia brought down by the rivers Isle and Dordogne. These alluvia tend to be more gravely in the upper sections, sandier in the lower zones to the west and south of the appellation. In the upper sections, the graves overflow into the commune of St.-Émilion around Château Figeac and Château Cheval Blanc.

The St.-Émilion Area

In St.-Émilion, the Merlot produces its finest results on the *Calcaire à Astéries*, the formation that supports most of the St.-Émilion Grands Crus Classés, at the top of the scarp overlooking the Dordogne Plain. The wines, often blended from the Merlot and Cabernet Franc, are powerful and generous with tremendous tannins. Professor Gérard Seguin from the University of Bordeaux has demonstrated that their quality is mainly determined by a consistent hydric regime. In effect, although these surface soils are at risk from drought in summer, the very compact *Calcaire à Astéries* lies close to the surface and can supply the vines with large quantities of water

through capillary migration (depending on their requirements during the ripening phase). On Fronsadais marls more or less covered with limestone fragments, the Merlot produces less structured wines with more supple tannins.

The Pomerol Area

Time for a change of terroir here, and a change in the characteristics of the Merlot wines. The "feminine" grape reigns supreme in a range of terroirs, all variously dominated by alluvial deposits from the Isle and Dordogne. These take the form of more or less argillaceous gravels in the upper sections, and sandy gravels or just sands lower down. In one particular area – and what an area – the alluvial deposits have disappeared altogether where the clay pushes through the gravel cover: We refer of course to the famous *boutonnière* (buttonhole) of Petrus.

These terroirs with their particular hydric regime bring out a wide range of aromas in the Merlot. The wines are powerful and astonishingly rounded on the palate, mainly due to their dense but exceptionally velvety tannins. Merlot wines from the more gravely terroirs surrounded by the lenticular clay strata are more robust with firm, rather more severe tannins but great depth of power and elegance. The wines from the sandy soils closer to the Isle and Dordogne, are less powerful but punchy all the same, with plenty of early-drinking quaffability.

The Merlot is less dominant in the rest of the Gironde. In the Médoc and Graves areas, it shares the terrain with the Cabernet Sauvignon, which tends to occupy gravelly mounds, while the Merlot keeps to argillaceous-calcareous soils. The Merlot continues to expand in the Bergerac viticultural area, where it currently accounts for more than half of all plantings destined for red wine production (Bergerac, Côtes-de-Bergerac, and Pécharmant). By contrast, it has lost ground in the Côtes-du-Marmandais and the Côtes-de-Duras, where it now represents just one third of planting acreage, almost equal with the two Cabernets (Franc and Sauvignon). The Buzet appellation, further up the river Garonne, is supported by the same geological formations found in the Monbazillac AOC, plus gravelly alluvia from the Garonne. The Cabernet Sauvignon prefers the gravels and the Cabernet Franc favors the boulbènes*, while the Merlot, which accounts for nearly half of the planting acreage, finds its voice in a wide variety of argillaceous-calcareous terroirs, ranging from limestone-free *terreforts* (semi-clayish/semi-chalky soils) to molasse and Tertiary limestone.

The Merlot Fashion

The Merlot is fairly widespread throughout the rest of southwest France, with the exception of the Béarn. It has soared in the Languedoc-Roussillon over the past 30 years, steadily replacing the low-quality, high-yield varieties. Agreeable and moderately aromatic, Merlot wines have a roundness that contrasts sharply with the rustic nature of some Mediterranean grape varieties. In these warm conditions however, the Merlot planted in more fertile soils can also yield potently alcoholic wines with a tendency to flabbiness. It gives much better results in cooler microclimates, such as in the Cabardès AOC or the Côtes-de-la-Malepère AQVDQS, where the Atlantic influence meets the Mediterranean weather system and the soils grow deep and calcareous once again.

In the rest of southern Europe, the Merlot is mainly established in Italy (where it produces remarkable results in the Trentino-Alto Adige, the Friuli-Venezia Giulia, and also in Tuscany) and it dominates plantings in the granitic terrain of Tessin, in southern Switzerland. In Eastern Europe, the Merlot is widely planted in Slovenia, Croatia, Bulgaria, Rumania, and Moldavia. In the USA, the Merlot grows principally in California, especially in the AVAs of Stags Leap and Russian River Valley.

The Appiano vineyard in the Trentino-Alto Adige, Italy.

The Tessin canton, Switzerland's fourth most important viticultural area. The Merlot accounts for 88 per cent of planting acreage and thrives in the near-Mediterranean climate.

The Jurançon and Béarn AOCs

The gentle croupes of the Jurançon AOC suit the Petit Manseng.

Along the northern edge of the Pyrenees, the ubiquitous vineyards of the Languedoc are replaced by small, scattered pockets of vineyards to the west. Gone are the thousand-hectare vineyards of the Aude and the Pyrénées-Orientales. Between them, the AOCs of Irouléguy, Jurançon, Madiran, and Béarn cover slightly more than 2,000 hectares while the AOVDQS appellations of Tursan and Côtes-de-St.-Mont, next to the Armagnac vineyards, extend for just 1,200 hectares.

In this area, the Pyrenean uplift had a direct impact on the viticultural soils. Regional morphology and geological formation were influenced by two fundamental stages in the uplift:

• The compression of the European plate and the Iberian mini plate, caught between the European plate and the African plate;

• The spread of a vast sheet of deposits produced by the intense erosion of the young, uplifting range.

Three Parallel Bands

The first stage formed part of a complex process involving the entire Pyrenean Mountain range, from the Atlantic to the Mediterranean. Running right across the northern side of the mountains are three east-west-facing parallel bands divided by faults. From south to north, you first have the axial zone where the mountains reach their highest altitude; then the intermediary northern-Pyrenean zone; then the lowest sub-Pyrenean zone.

The northern-Pyrenean zone is actually the lowest part of the Aquitaine Basin, that accumulated considerable layers of deposits throughout the Mesozoic Era. Held in a vise-like grip between the axial zone and the sub-Pyrenean zone, these Jurassic and Cretaceous layers were tightly folded, then tilted northward, straddling the sub-Pyrenean zone. The group of layers was then eroded, exposing the core of the folds at the surface.

The second stage began in the Eocene, at the start of the Tertiary Period. Erosional materials from the young, blossoming range immediately started to accumulate at the foot of the Pyrenees (Palassou pudding stone*), and then gradually covered a wider and wider area as uplift intensified.

In the Miocene, the debris stripped from the highest mountain peaks spread in a vast fan to form the immense Lannemezan Cone. The Garonne River, on its course to the Atlantic, was forced eastward around the Cone, via Toulouse and Agen. The Jurançon pudding stone was part of this Miocene formation. Toward the end of the Tertiary, this formation would be covered by the clays and gravels that extend all the way to

3 800	2 500	Millions of years	540	500	435	410	360

Archaen	Proterozoic	Cambrian	Orodovician	Silurian	Devonian	Carbonif
	PRECAMBRIAN				PALEOZOIC	

the center of the Aquitaine Basin, with the finer elements being carried the furthest. Throughout the Quaternary Period, the torrents that flowed down from the Pyrenees continued to spread vast expanses of fill (Villafranchian, then Quaternary terraces) along the length of their banks. When the entire mountain range, including the northern border, underwent the final period of uplift, the torrents and rivers (*gaves* is the local name for streams originating in the mountains) became boxed in by the formations described above. In the process, the Tertiary and Quaternary deposits were formed into strips of terrain running alongside the water.

Erosion Good for Viticulture

While all quite varied, the Béarne vineyards (Béarn-Bellocq, Jurançon, Madiran and the regional Béarne appellation) are directly related to the Pyrenean uplift. The Bellocq vineyard is established on the overthrust zone around Salies-de-Béarn and Orthez; the Miocene deposition of pebbles was restricted to the northern zone that is also overlain by a gravelly Würm glacial terrace along the Gave of Pau.

In the Jurançon area, the Mesozoic overthrust zone is much less obvious since it is largely overlain by erosional products: Palassou pudding stone, then Jurançon pudding stone, gravelly clays, then Late Tertiary Villafranchian nappes. The soils derived from these very gravelly formations share similar characteristics. They are quick to warm up and well drained – a most

important feature in a region with the highest rainfall in France (47-51 inches per year), which favors the production of dessert wines. Jurançon dessert wines have a characteristically honeyed bouquet with notes of cinnamon, nutmeg and candied fruit. They are refreshing and generous and typical of the Petit Manseng grape, which is in its element in this terroir.

The Madiran vineyard occupies formations dating from the same era, although the constituent elements are finer as you go deeper into the basin. Here, the Miocene molasse of Armagnac corresponds to the Jurançon pudding stone. Gravelly clays and the Villafranchian form the upper slopes, slashed by rivers into long, generally north-south facing ridges. The Tannat produces its best results in this terroir, yielding raspberry-scented young wines that, with age, develop spicy, toasty aromas. Their rugged tannins mellow over time, giving the wines strength and silky, mature body.

Madiran and Pacherenc-du-Vic-Bilh appellation wines originate from the same area: Armagnac molasse overlain by gravelly clays and Villafranchian scree.

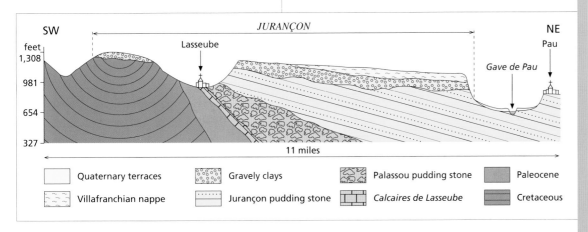

Cross section, Jurançon plateau.

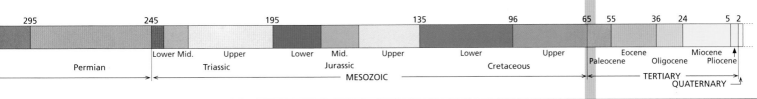

Jerez and the Vineyards of Andalusia

The geological history of Spain is inseparable from that of Portugal. The Iberian block forms part of what geologists call a "craton," meaning a stable area of continental crust. It evolved from repeated orogenesis that did not affect the separate, European craton.

At the end of the Paleozoic Era, the Iberian craton was a long way from its present location. The Gulf of Gascony did not exist and the hills of Galicea were near the Massif Central. The Iberian block was above water and would remain so for most of the Mesozoic Era and Tertiary Period. The Jurassic marked the start of a vast process of drift and rotation that would eventually place the Iberian Peninsula where it is today.

When Two Blocks Collide

The Iberian block, squashed between the west European block and the African block, collided with the west European block to the north. This was the collision that led to the formation of the Pyrenees. Another Miocene collision, in the south between the African plate and the Iberian craton, led to the formation of the Rif and Atlas ranges in North Africa and to the uplift of the Betic Mountain belt and the Balearic Islands in Spain. The old Iberian block, trapped in a chain of collisions on either side, was compressed between Europe and Africa. Rigid though it was, the surface of the block buckled, creating huge ripples across the Earth's crust, sometimes bordered by faults. These were the origins of the inland ranges, the Catalan Mountains, the Iberian range, and the Central Mountains, where the Hercynian formations burst through the overlying Mesozoic layers.

Two Mountain Ranges

Today, the Spanish landscape is essentially mountainous with two major ranges: in the north, the Pyrenees (rising to 11,130 feet in the Pico de Aneto); and in the south, the Betic mountain belt that includes the Sierra Nevada, home of Spain's highest peak (11,373 feet). The area between consists of the following:
• Two high plateaus of Tertiary fill, separated by the Central Mountains (Sierra de Gredos and Sierra de Guadarrama): the northern Meseta in the northwest (Castilla-Léon); and the southern Meseta in the east-central region (Castilla-La Mancha).
• Two depressions: the Andalusian Plain in the south (Guadalquivir basin) between the Betic belt and Sierra Morena; and the Ebre basin between the Pyrenees and the Iberian range.

The rolling vineyards of Jerez are located around Jerez de la Frontera and Sanlúcar de Barrameda.

3 800	2 500	Millions of years		540	500		435	410		360
	Archaen		Proterozoic		Cambrian	Orodovician		Silurian	Devonian	Carbonife

PRECAMBRIAN — PALEOZOIC

Andalucia is home to Spain's only real coastal plain, crossed by the Gualdalquivir River on the last, lazy leg of its journey from Seville to Sanlúcar de Barrameda. This is the vast Guadalquivir Valley Basin, an east-northeast, west-southwest-facing depression that encompasses the entire region, squeezed between the Sierra Morena in the north and the Betic belt in the south. The Sierra Morena range rises no higher than 4,251 feet in the Sierra Madrona to the north of Cordoba, whereas the Betic belt boasts some of the highest peaks in the Iberian Peninsula (such as the 11,373-foot high Cerro Mulhacén in the Sierra Nevada). The landscape here is particularly rugged, with ranges (Sierras de Segura, de Baza, and Las Estancias) facing in all directions that chop up the massif in a disorderly fashion, as if bearing witness to their turbulent geological past.

The Betic Mountain Range

This is an Upper Tertiary Alpine range, uplifted when the African plate and the Iberian craton collided. Part of the strata laid down from the Paleozoic to the Tertiary were violently compressed and folded by the force of the impact, thrusting upward and tilting northward to form the Betic system. On the African continent, meanwhile, the strata were tilted southward, forming the Rif in northern Morocco and the Tellian Atlas Mountains that tower over the Mediterranean coast in Algeria. The Oligocene witnessed, on the one hand, the collapse of the valley that is now occupied by the Guadalquivir River, and on the other hand, the formation of a trough region in Andalusia that emerged from underwater. On the borders of the uplifting Betic Mountains, this trough gradually filled up, first with marine sediments (in the Oligocene when the trough was underwater) then with Miocene peri-Alpine molasse washed down by the Gualdalquivir and its tributaries in the Pliocene and the Quaternary. The alluvial deposits spread by the rivers filled the entire coastal plain (Las Marismas). What we see today is a vast U-shaped region opening to the Atlantic, whose southern arm forms a massive, almost impenetrable barrier that suddenly drops down to the sea.

The Andalusian Terroir

The climate here is predominantly Mediterranean, with Atlantic weather influences on the coastal plain along the Gulf of Cadiz. Broadly speaking, the eastern part is hot and dry, with mean annual temperatures as high as 63° F. In July and August, the thermometer often rises to 104° F, and there is sometimes no rain at all in summer, with the east winds aggravating the drought conditions. On the coastal plain however, it rains heavily in the autumn and spring thanks to weather influences from the Atlantic. So despite the summer drought, the Jerez region as a whole has a mean annual rainfall of just over 23 inches.

The Andalusian vineyards may be divided into four main regions: Jerez, on the coastal plain, between the mouths of the Guadalquivir and Gaudalete rivers; *Condado de Huelva*, slightly farther west; *Montilla-Moriles* in the foothills of the Betic range, south of Cordoba; and the *Malaga* region, with the city at its center.

The *Albarizas* of Jerez

One of the special features of Jerez wines, or sherry, is the aging technique, under a veil of yeasts called the "flor" (flower) that inhibits oxidation and gives the wines their characteristically nutty, tangy bouquet of walnuts mixed with over-ripe apples and roasted coffee beans. Fino, Amontillado and Oloroso sherries are especially distinctive. The first are dry and delicate and fortified with grape spirit to 15 per cent alcohol, allowing the *flor* to grow on the surface of the wine. Sanlúcar de Barameda Fino sherries are called *Manzanillas* (meaning apples). Amontillado sherries are aged according to a more complex process and fortified in the course of fermentation with a higher level of alcohol (17 per cent) that kills off the *flor*. The wines are then aged according to the more complex *Solera* system that essentially consists of blending younger wines with older ones across a series of casks (called the solera). Oloroso sherries are fortified with 18 per cent alcohol that inhibits development of the flor, then aged for up to 10 years in the wood.

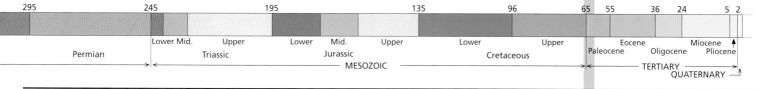

295		245		195		135		96	65	55	36	24	5 2

Permian | Lower Mid. Upper | Lower Mid. Upper | Lower Upper | Paleocene | Eocene Oligocene | Miocene Pliocene

Triassic | Jurassic | Cretaceous

MESOZOIC | TERTIARY / QUATERNARY

The Jerez terroir forms a precisely limited area bordered by three main towns: Sanlúcar de Barrameda, at the mouth of the Guadalquivir; Puerto de Santa Maria, at the mouth of the Guadelete River; and especially Jerez de la Frontera, some 12 miles inland. There are three types of soils in the sherry-producing area, *albarizas*, *barros* and *arenas*, which differ in the richness of their calcium content.

The best terroirs are those supported by the *albarizas*. These are white, Upper Oligocene marls that originated in the deep, tranquil waters of the previously submerged trough of

Distribution area of the albarizas and barros soils around Jerez de la Frontera and Sanlúcar de Barrameda.

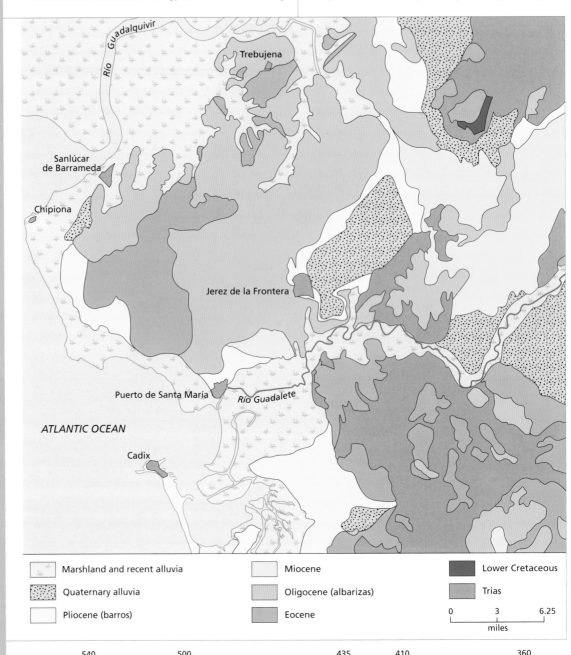

	Marshland and recent alluvia		Miocene		Lower Cretaceous
	Quaternary alluvia		Oligocene (albarizas)		Trias
	Pliocene (barros)		Eocene		

0   3   6.25
miles

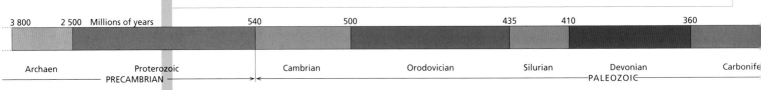

The albarizas are to Jerez what the gravels are to the Médoc or the Kimmeridgian marls are to Chablis.

Andalusia. These legendary soils are to Jerez what the gravels are to the Médoc, or the Kimmeridgian marls are to Chablis. The *albarizas* were formed by the sedimentation of marine diatoms and radiolarians, microscopic algae with siliceous shells that then mixed with fairly fine-grained sand. By some extraordinary stroke of luck, these marls are rich in Montmorillonite clay, a mineral that absorbs water throughout the rainy period in the spring, then provides the vines with a steady supply of water in summer. These soils are perfectly adapted to the Palomino, source of the finest sherry and Jerez' premier grape accounting for more than 90 per cent of plantings. You can see why only vineyards supported by the Oligocene white marl formations are officially entitled to produce Jerez Superiores wines.

The *albarizas* form the upper section of two series of *croupes* that run parallel to the Guadalquivir: one extends from Puerto de Santa Maria, to the north of the town of Jerez; the second stretches from the south of Sanlúcar de Barrameda to the north of the Trebujena wine-producing area.

The *barros* and *arenas* soils are sandier and less calcium-rich than the *albarizas* and of more varied geological origin (Tertiary and Quaternary). They are mainly found along the coastline from Sanlúcar de Barrameda and Chipiona as far as Puerto de Santa Maria, and also in northeast Jerez. The Pedro Ximénez grape is more at home here than the Palomino.

Montilla-Moriles *Albaros* and *Ruedas*

We find the same soil types in the Montilla-Moriles appellation, a producer that for many years was overshadowed by its more illustrious neighbor, Jerez. More recent and less well-known, Montilla-Moriles traditionally supplied Jerez with large quantities of single varietal Pedro Ximénez wines. The rolling vineyards of Montilla-Moriles, planted with the Pedro Ximénez, form a red and white landscape of white soils known locally as *albaros*, and red soils called *ruedas*. The *albaros* are identical to the Jerez *albarizas* whereas the *ruedas* are sandier and less calcareous. The Pedro Ximénez yields the same range of products as in Jerez although there is more of an emphasis on sweet wines. Pedro Ximénez wines are made from grapes that have been raisined to increase sugar concentrations. Fermentation is arrested by *mutage* (chemical sterilization) so as to retain a sense of creaminess that offsets the natural fruit acidity.

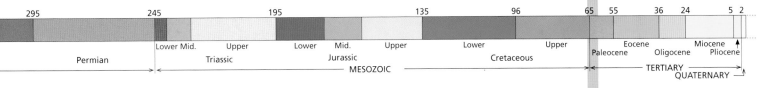

295	245	195	135	96	65	55	36	24	5	2

| Permian | Lower Mid. | Upper Triassic | Lower | Mid. Jurassic | Upper | Lower | Upper Cretaceous | Paleocene | Eocene | Oligocene | Miocene Pliocene |

MESOZOIC — TERTIARY — QUATERNARY

The Duero Valley

The vineyards of the Spanish Duero region (Douro), like those of La Rioja and Navarre in the Ebre Valley, occupy a vast sedimentary basin filled with Tertiary formations and drained by one of the biggest rivers on the Iberian Peninsula. There, however, the comparison stops, since there are fundamental differences between the two regions.

The Ebre Basin faces east toward the Mediterranean whereas the Duero opens to the Altantic Ocean and Portugal, where it is better known as the Douro River, famous for its spectacular handiwork carving out the vineyard slopes of the Port region (*cf.* p. 182). The rivers are also located at different altitudes: the Ebre runs at a maximum of 1,600 feet in Conchas de Haro in La Rioja region, but drops to 850 feet in Tudela in Navarre; the Duero only drops below 1,600 feet on its way out of Spain into Portugal. As a result, most of the big wine-producing zones in Castilla-Léon (Duero basin) are located at heights of 1,900-2,600 feet, with some vines on the rim of the basin located at more than 3,270 feet.

The red wine-producing vineyards of Toro, a Denomination of Origin area in the province of Zamora.

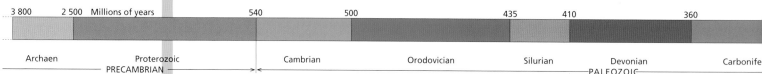

3 800	2 500	Millions of years	540	500	435	410	360

Archaen	Proterozoic	Cambrian	Orodovician	Silurian	Devonian	Carbonife

PRECAMBRIAN — PALEOZOIC

The scenery around Peñafiel (monastery of San Pablo) in the Ribera del Duero region.

A Vast Expanse of High Plateau

The Duero Basin, part of the northern Meseta or Castilla la Vieja, is a high plateau with extreme weather conditions. Cold winters and spring frosts, often with devastating consequences for the vines, are followed by hot, dry summers mitigated only by cool nights that encourage ripening and optimum color development. The region is currently enjoying a revival and boasts four "Denomination of Origin" (DO) areas: Toro, Rueda, Cigales and especially Ribera del Duero.

Its geological history is fairly straightforward. Castilla-Léon is located in the northern part of the Meseta, a vast, compact plateau created by successive periods of orogenesis dating from the Pre-Hercynian to the Hercynian. This old, flattish block, worn away by erosion, occupies the northwestern half of the Iberian Peninsula. It remained above water from the end of the Paleozoic to the end of the Mesozoic eras, since the Jurassic and Cretaceous seas barely made it to the eastern rim of the basin (Cantabrican Mountains and Iberian range) around Burgos and Soria.

At the beginning of the Tertiary, this region was a huge closed basin that became filled with continental deposits, particularly Eocene then Miocene sediments, that today cover most of the Spanish Duero basin. More than 800 feet thick, these Miocene formations consist of two types of deposits:

• Soft sediments: Tortonian sands and argillaceous sands, capped by grayish marls, sands and Sarmatian* gypsum. They fill the entire base of the basin, gradually thinning out toward the west where they directly overlay the Paleozoic basement, west of a line running from Zamora to Salamanca.

• Hard sediments: lacustrine Pontian limestone. In the northeast sector only, these overlay the preceding formations within a triangle running from Burgos to Soria and the town of Valladolid, north of which they come to a point in the Torozos Mountains.

The Duero, Sculptor of Landscapes

At the end of the Tertiary, the Cantabrian Mountains and the Iberian Range received their final uplift in the last convulsions of the Pyrenean-Alpine folding. This process had two outcomes. The first, in the north, was the uplifting of the Cantabrian Mountains, consisting here of gneiss and carboniferous quartzite, along an east-west fault running from

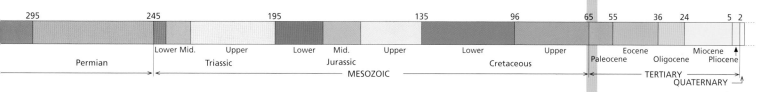

295		245		195			135		96		65	55		36	24		5	2
		Lower	Mid.	Upper	Lower	Mid.	Upper		Lower	Upper		Eocene			Miocene		
Permian			Triassic			Jurassic			Cretaceous		Paleocene		Oligocene		Pliocene		
						MESOZOIC						TERTIARY					
													QUATERNARY				

Peñafiel Castle keeps watch over the austere landscape of the Ribera del Duero region.

3 800	2 500	Millions of years		540	500		435	410		360
Archaen	Proterozoic			Cambrian	Orodovician		Silurian	Devonian		Carbonife
	PRECAMBRIAN							PALEOZOIC		

La Robla to Cuevere. In the Pliocene, erosion stripped these mountains of vast quantities of debris that covered the soft Miocene strata throughout the northwest part of the basin. The second outcome was the tilting of the basin toward the west that forced the Duero to change direction and flow toward the Atlantic. The river and its tributaries, especially the Pisuerga that comes from the north, became entrenched in the Miocene formations. In the northeast, they sheared through the hard, straight Pontian calcium strata, opening up valleys in the softer subjacent formations (Tortonian-Samatian). Between these valleys, some wider than others, vast, sharp-edged plateaus called *páramos* were formed. In the northwest, the rivers that descended from the Cantabrian Mountains (Cea, Elsa, Orbigo) dug valleys in Pliocene gravelly deposits, forming another type of *páramos* with less abrupt edges. In the rest of the Miocene basin, vast plains of clays and sands on either side of the Duero Valley were laid bare by erosion, creating a landscape of pine-covered dunes between the river and Segovia.

The river remained at high altitude in the Miocene formations but once past Zamora (2,090 feet) it gouged out the schist, gneiss, and granite of the Paleozoic basement, carving out the sheer-sided gorges along the frontier with Portugal that herald the spectacular port vineyards farther downstream.

Ribera del Duero

Dedicated to red and rosé wine production, the Ribera del Duero DO area spreads along the banks of the Duero from San Estebán de Gormez to Valbuena de Duero. The river here carved through the lacustrine limestone, cutting a valley of variable width through the *páramos*. Narrow in San Estebán de Gormez (Soria province), it gradually broadens toward Roa in Burgos province, then squeezes through a tight gap between two *páramos* before opening out once again around Peñafiel and Valbueno de Duero. Flanked by scrub oaks and a few cereal crops on the *páramos*, the vines are to be found in the soft Miocene formations, the argillaceous-stony soils of the valley sides (Tortonian) and the terraces along the river.

This tough terrain suits the Tinta del País, a grape variety similar to the Tempranillo. Despite the extreme weather conditions, it develops good color thanks to the particular combination of hot days and cool nights in the ripening period. The wines produced are dark red, powerful and robust, and require no blending with other varietals. A few decades ago, plantings were mainly focused on Peñafiel but these days Tinta del País is increasingly cultivated farther east, around Aranda del Duero. This is the home of two of the most outstanding Spanish wines: Viña Pesquera, a single varietal Tinta del País wine; and Vega Sicilia, the older of the two, blended from the Tinta del País with Merlot and Cabernet. Plantings of these two Bordeaux varieties are now strictly regulated by the rules of Denomination of Origin.

Rueda and Toro

Downstream, the vineyards switch from red to white in the Tierra de Medina area. This is the home of the Rueda (DO) vineyards where the Verdejo flourishes in the upper gravely terraces along the water's edge, and in some of the argillaceous terrain. Rueda produces dry and dessert white wines and sherry-type wines.

Farther along the Duero, we find a return to red wine production in the Toro (DO) region. The vines continue to occupy the upper gravelly terraces flanked by now-irrigated alluvium and the scrubby plateau. To the north of Valladolid, along the Pisuerga, a tributary of the Duero, a broad plain opens out between the *páramos*. This is the location of the Cigales (DO) area, famous for its rosé wine based on the Tinta del País and the Garnacha blended with white varieties. The vines, as before, remain on the upper gravelly terraces along the river.

South of Léon, non-DO wines are produced from the gravelly terrain on the right-bank of the Esla (Valdevimbre vineyards) and from the sheer, terraced slopes of the *arribas*. This is an especially precipitous part of the Duero with a near-Mediterranean climate that produces agreeable clarets from the Juan García grape.

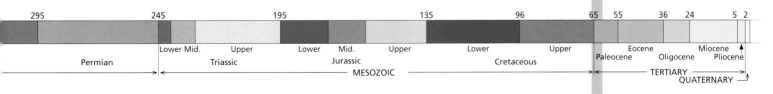

The Mediterranean Basin

The Mediterranean Basin has one of the most complex geological histories in the whole of France. Some viticultural areas, such as the Minervois and Faugères appellations, are confined to certain selected rock types, whereas others, such as Corbières or Coteaux-du-Languedoc, span the entire series of geological formations, from the Paleozoic to the present-day.

The origins of Mediterranean soils lie partly in the Paleozoic Era. Successive marine deposits (Ordovician schists, Devonian limestones, Carboniferous schists) were modified in the course of the Hercynian orogenesis, and were sometimes metamorphosed and intruded by granite. The basement rock became exposed in the Maures Massif, the Mouthoumet Massif (Corbières), and Faugères, then eroded at the end of the Paleozoic Era under the tropical weather conditions. In the process, a depression formed around the Maures Massif, dug out of soft Permian sandstone and the rich, red sandstone of Lodève (ruffes*).

The Aude Valley between Alet-les-Bains and Couiza.

Zones Above Water

At the beginning of the Mesozoic, there were three main zones above water: in the south, the Pyrenean-Provençal axis that included the Corsica-Sardinian landmass; in the northwest, the Montagne Noire complex that included the Mouthoumet Massif; and in the northeast, the Cévennes Mountains. In the Jurassic, the waters of the Alpine Sea invaded from the east.

The vineyards around Limoux cover extensive slopes of Lutetian molasse shed by the uplifting Pyrenees.

3 800	2 500	Millions of years	540	500	435	410	360
Archaen	Proterozoic		Cambrian	Orodovician	Silurian	Devonian	Carbonife
	PRECAMBRIAN					PALEOZOIC	

*The Côtes-de-Provence
AOC, around
St.-André-de-Figuière.*

In order to reach the Aquitaine basin, they were forced to skirt the Montagne Noire complex on its northern (Grands Causses formation of the Massif Central) and southern sides (north Pyrenean zone, south of the Limousin and Les Corbières). Things remained largely unchanged in the Lower Cretaceous, except that the sea retreated from the Grands Causses; also the garrigues (limestone scrubland) around Nîmes were formed.

The Pyrenean-Provençal Uplift

The Mid-Cretaceous brought the first convulsions of the Pyrenean uplift and the formation of huge east-west undulations filled with bauxite* deposits that developed in the humid, tropical climate. This occurred to the west of Montpellier and in the Durance isthmus between the Cévennes Mountains and the Maures Massif. In the Upper Cretaceous, a marine incursion from the newly formed Atlantic Ocean created a long gulf in the north Pyrenean zone. It extended all the way to Provence, now separated from the Alpine Sea by the Durance isthmus.

As the Pyrenean uplift continued into the Lower Tertiary (Eocene), another invasion from the Atlantic was driven northward. The waters reached as far inland as the eastern edge of St. Chinian, submerging the north of Les Corbières and the Minerve region (foraminiferal* limestone deposits) for the first time since the Paleozoic Era.

The Pyrenean-Provençal uplift rose to a crescendo in the Upper Eocene, by which time the entire Languedoc region was above water. With each successive thrust, the Mesozoic layers of the north Pyrenean zone were folded then tilted violently northward onto the sub-Pyrenean zone. A simple recumbent fold south of the Limousin was part of a long and intricate movement that sent a huge nappe thrusting miles into the eastern Corbières. Narrowly missing the Montagne Noire around St.-Chinian, it then headed north of Montpellier and plunged headlong beneath the Camargue. Bits of this nappe moved farther north, creating the rigid limestone fold of the Pic St.-Loup Mountain. The area north of Marseilles and the Massif des Maures was also caught up in this process, during which nappes of Triassic, Jurassic, and Cretaceous strata were folded, fractured, and sometimes overturned. Now the aftershocks of the Alpine Storm would make themselves felt.

Collapsed Basins

In the Oligocene, collapsed basins (Alès, Sommières) formed along old scars in the Hercynian basement (southwest, northeast faults) that filled up at the same rate as they sank. The Miocene marked a cataclysm that changed the face of the entire region: the collapse of the center of the Pyrenean-Provençal range together with the opening-up of the Golfe du Lion. The Languedoc, like Provence, now tilted southward – facing the Mediterranean, in other words. In the Mid-Miocene (Helvitian-Burdigalian), the lower Languedoc as far as the foot of the Montagne Noire was once again under water. As the Mediterranean dried up at the end of the Miocene, river valleys became steeper (gorges of the Orb and Hérault rivers). In the Pliocene, the sea launched a final invasion into the steadily sinking lower Languedoc. In the Quaternary, a series of gravel terraces was formed, on the boundaries of bodies of water.

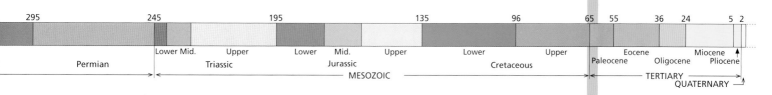

The Limousin and Cabardès

The terroirs of the Limoux and Cabardès AOCs, in the easternmost part of the region, do not fit the conventional profile of the other Languedoc-Roussillon vineyards. Situated between the Atlantic Ocean and the Mediterranean Sea, open to the tempering influences of the Atlantic Ocean, they are unusual in several ways including altitude and aspect.

Arrangement of Tertiary

layers overlaying

the ancient basement

in Cabardès and

the Limousin.

The Limoux AOC falls entirely within the sub-Pyrenean zone while the Cabardès AOC occupies the southern side of the Montagne Noire. The region remained above water practically throughout the Mesozoic Era. With the exception of the Campanian Alet sandstone formation, the bulk of deposits are from the Tertiary Period. The foundations of the Limoux, La Malepère, and Cabardès vineyards were in fact laid over a very short period of time (Paleocene-Eocene) thanks to three main geological processes.

A Three Stage Formation

The first process forms part of a series of Lower Eocene marine invasions that submerged the entire region from the area north of Cabardès to the south of Limoux. The engulfing seas left alternating layers of hard and soft deposits: sandy limestone (Thanatian), red sandy clays (Spanacian), hard foraminiferal*

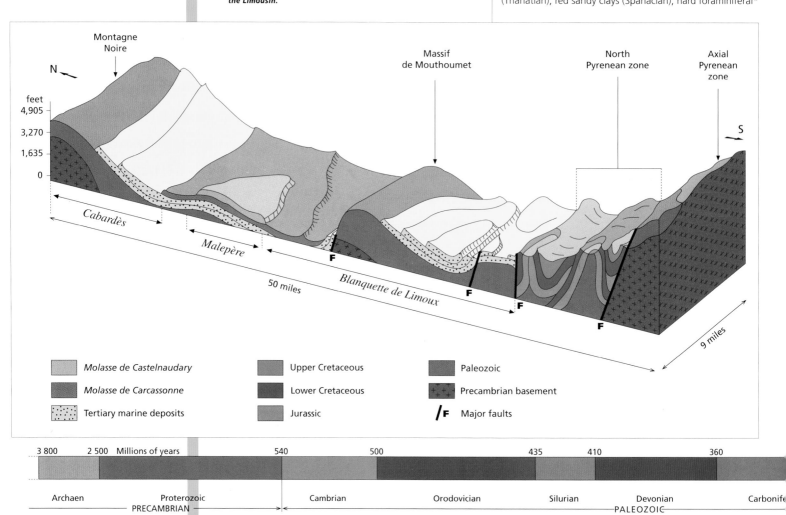

Legend:

- Molasse de Castelnaudary
- Molasse de Carcassonne
- Tertiary marine deposits
- Upper Cretaceous
- Lower Cretaceous
- Jurassic
- Paleozoic
- Precambrian basement
- F Major faults

The ruined castle of Coustaussa, near Couiza in the Sals Valley, with the mysterious site of Rennes-le-Château looming up in the background.

limestone, blue Turritellae*-bearing marls and oyster-bearing sandstone banks (Ilerdian).

The second process is linked to the Pyrenean uplift in the Mid-Eocene. The above deposits were largely overlain by accumulations of Lutetian detritus that was stripped from the new peaks and washed down by rivers. The third process took place in the Upper Eocene when the intense compression between the Pyrenees and the Montagne Noire led to the uplift of the Massif de Mouthoumet. The southern portion of the Limoux appellation was formed by a single block that tilted up rather like a drawbridge, with the highest part being in the center of the region (the 2,100-foot Pic de Brau to the west of the Aude River and the 2,960-foot Pic de Fondondy to the east). This had a dramatic impact on the landscape and the vineyards, leaving an east-west fault with a ledge more than 1,600 feet high that towers over the town of Limoux.

Vineyards Supported by Soft Formations

In the southern half of the Limoux appellation, the hills range from 980-2,600 feet (the highest vineyards reaching more than 1,600 feet in Conilhac-de-la-Montagne) and the strata slope sharply southward. The sedimentary layers covering the basement rock have been largely eroded, leaving the bedrock exposed but vine-free to the east of the Alet-les-Bains defile. In addition to wearing away practically all of the Lutetian detritus, erosion has also exposed the hard layers (Thanetian limestone, foraminiferal* limestone, and oyster-bearing sandstone banks) forming the cuestas* that give the scenery such a rugged look. The vineyards are mainly established in small basins dug out of the softer formations: red Spanacian marls in Luc-sur-Aude; blue Turritellae*-bearing marls in the Couiza-Coustaussa valley in Roquetaillade; and red Maestrichtian marls in Campagne-sur-Aude.

Molasse de Carcassonne

Altitudes in the area around Limoux, to the north of the large fault that runs along the sheer flank of the Massif de Mouthoumet, are much lower (460-1,300 feet) and the topography is completely different. Here, the strata slope northward. The vineyards are established in Lutetian detrital molasse deposited in two stages that corresponded to two main thrusts of the Pyrenean uplift. Between the two came a period of calm when the lacustrine limestone was deposited, now forming a cuesta, which extends from Castelreng to Toureilles, Limoux and St.-Hilaire.

The cuesta topography in this area was also produced by

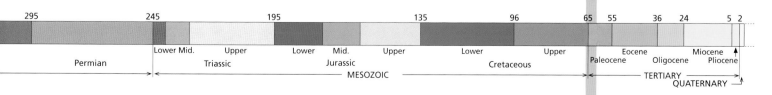

295		245		195			135		96		65	55		36	24		5	2
		Lower Mid.	Upper		Lower	Mid.	Upper		Lower	Upper			Eocene		Miocene		
Permian		Triassic			Jurassic				Cretaceous		Paleocene		Oligocene			Pliocene	
						MESOZOIC							TERTIARY				
													QUATERNARY				

Above:

the vineyards of

Campagne-sur-Aude,

planted in red

Maestrichtian marls

(Upper Cretaceous).

Towering in the

background to the west

is the Pech de Bugarech,

the highest peak in the

Corbières Mountains.

Right:

the village of

Roquetaillade built on

Ilerdian foraminiferal*

limestone. Most of the

vineyards are located on

blue Turritellae-bearing

marls capped by oyster-

bearing sandstone

banks that form the

ridge of the cuestas.

erosion, except that here the cuestas are reversed. They are eroded out of *Molasse de Carcassonne*, a formation consisting of alternating sandstone banks and marly beds that support all the vineyards of the northern sector, entirely located in the principal basins between the cuestas.

Viticultural Terroirs Delimited by Climate

The vineyards of the Limoux AOC, are exclusively supported by limestone soils, and are predominantly planted to the Mauzac plus, from 1970 onward, the Chenin and especially the Chardonnay. The range of wines includes one of the oldest in France, Blanquette Méthode Ancestrale, a naturally effervescent wine exclusively based on the Mauzac and containing no added liqueur. So as to ensure optimum sweetness, the berries are picked when extremely ripe from vines planted in sheltered positions. The cradle of production lies to the west of Limoux, in the molassic basin of Magrie and de Toureilles, an ideally sheltered location tucked between the uplifted block of the Massif de Mouthoumet in the south, and the Lutetian lacustrine limestone in the north.

Given the relatively homogeneous soils, the characteristics of Blanquette-de-Limoux and Crémant-de-Limoux very much depend on the location of the terroir and especially the local climate. The wines are robust from Limoux and St.-Hilaire in the north where the climate is Mediterranean, whereas those from the cooler upper Aude Valley are fine and elegant.

The Chardonnay has found a little niche in the Limousin where it develops a range of very unusual characteristics. Following a detailed survey of the appellation terroirs, the Limoux Wine Cooperative now produces four highly distinctive cuvées, each one illustrating the Chardonnay's markedly different response depending on the terroir of origin: Mediterranean, oceanic, Autun area, and upper valley.

The Cabardès AOC, to the north of Carcassone and west of the Minerve area, occupies the southern slopes of the Montagne Noire. The formations in this region are the same as in the Limoux AOC, essentially consisting of Lower Eocene marine limestone overlaying the schists and gneiss of the sparsely planted ancient bedrock. In the southeast, between Pennautier and Conques, these deposits are overlain by lacustrine *Calcaire de Ventenac* then by *Molasse de Carcassonne*. These stony, shallow limestone terroirs at the confluence of Mediterranean and Atlantic weather influences are familiar territory for the Grenache Noir and the Syrah, but they also suit the Merlot and the Cabernet Sauvignon that usually prefer deeper soils of marls and molasse. This blend of Bordeaux and Mediterranean varieties produces robust, elegant reds and rosés with an unmistakable strawberry fragrance. On the palate, the mellowness of the Merlot harmonizes beautifully with the tannic strength of the Syrah.

3 800	2 500	Millions of years	540	500	435	410	360

Archaen	Proterozoic	Cambrian	Orodovician	Silurian	Devonian	Carboni
PRECAMBRIAN				PALEOZOIC		

Alet-les-Bains is the gateway to the upper valley where the Aude River has carved narrow gorges out of the Massif de Mouthoument that towers over the central Limousin landscape.

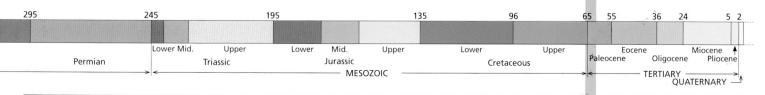

295	245	195	135	96	65	55	36	24	5	2

Permian	Lower Mid.	Upper	Lower	Mid.	Upper	Lower	Upper	Paleocene	Eocene	Oligocene	Miocene	Pliocene
	Triassic		Jurassic			Cretaceous						

MESOZOIC

TERTIARY

QUATERNARY

The Corbières Region

The Corbières region is a real hodgepodge from the geological point of view. This vast square some 40 miles on each side boasts a greater variety of soil foundations than any other French appellation, spanning practically the entire geological series, from the Paleozoic formations to the stony Quaternary terraces.

With soils from virtually every geological series and a variety of different microclimates, the Corbières region produces an exceptional range of wines. In addition to the Corbières AOC itself, there is Fitou, an appellation area comprising nine communes specializing in dry red wines, and the Roussillon AOC, famous for its Rivesaltes and Muscat-de-Rivesaltes *Vins Doux Naturels* (French term meaning "naturally sweet wine". Abbreviated to VDN.), that extends into the same communes as the Fitou AOC.

The geological jumble of the Corbières region.

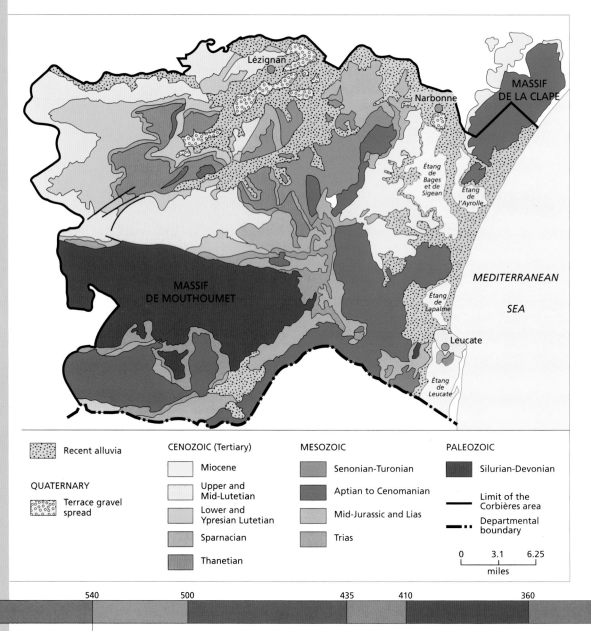

Recent alluvia

QUATERNARY

Terrace gravel spread

CENOZOIC (Tertiary)

Miocene

Upper and Mid-Lutetian

Lower and Ypresian Lutetian

Sparnacian

Thanetian

MESOZOIC

Senonian-Turonian

Aptian to Cenomanian

Mid-Jurassic and Lias

Trias

PALEOZOIC

Silurian-Devonian

Limit of the Corbières area

Departmental boundary

0   3.1   6.25
miles

3 800	2 500	Millions of years	540	500	435	410	360

Archaen	Proterozoic	Cambrian	Orodovician	Silurian	Devonian	Carbonif

PRECAMBRIAN ———— PALEOZOIC

Between the Montagne Noire and the Pyrenees

The Corbières appellation area is the eastern continuation of the Limoux AOC with which it has much in common. The entire Corbières region actually falls within the sub-Pyrenean zone whose southern boundary roughly coincides with the departmental boundary of the Aude (cf. the Roussillon area pp. 120-121). We find the same Lower Eocene marine limestone deposits and Lutetian erosional fill from the Pyrenees and the same uplift of the Massif de Mouthoumet along a large east-west fault, with more significant outcrops of basement rock due to erosion.

Despite these similarities, the layout of the Corbières AOC is actually quite unlike that of the Limousin, due to four localized geological phenomena. The first two arose from the close proximity of the Montagne Noire and the Pyrenees that here acted like the jaws of a vise.

The most remarkable phenomenon was the development of the nappe of the eastern Corbières. Accumulations of Triassic, Jurassic, and Cretaceous deposits in the north Pyrenean zone depression (corresponding to the north of Roussillon) were folded, compressed and thrust northward for miles, like a huge layer of icing slipping off a tilted cake. They spread over the eastern half of the region, and have now been reduced by erosion to a chaotic jumble of all the formations in the nappe plus the layers it covered.

The second phenomenon was the formation of the huge folds in the northern half of the Corbières, especially the folded mountain of the Mont-Alaric that burst through the covering of Lutetian molasse, revealing the Lower Eocene marine limestone in the upper part of the anticline. The third phenomenon, unknown in the Limousin, was the Miocene marine incursion in the Narbonne region that deposited more or less argillaceous sands. The fourth phenomenon is the extensive gravel terrace deposits that were spread in the Quaternary by the Aude and its tributaries, especially around Lézignan and Narbonne.

Corbières: Spanning Geological Time

The Corbières region is essentially a red-wine producing area predominantly planted to the Carignan together with the auxiliary varieties, Grenache Noir, Cinsault, and increasingly the Syrah or the Mourvèdre. With vineyards founded on rocks of every geological series, Corbières produces a wide range of wines that vary according to subsoil and local climate. Starting in the Paleozoic Era in Villeneuve-lès-Corbières, Cascatel, and north of Tuchan: the vines here are supported by ancient schists and a cool climate due to the altitude, yielding elegant, delicate wines. Moving up the geological series, we find red Triassic marls in St.-Jean-de-Barrou; Liassic marly limestone in the coastal rim of the Fitou region; Lower Cretaceous Urgonian limestone between the coastal sector and the Durban region, home to robust, powerful wines originating in scree on the fringes; and Senonian reef limestone in the Fontfroide region. In the northwest of the appellation, the Lutetian formation in the Dagne Valley (gray marls and limestone banks) only suits the Syrah, whereas neither the terroir nor the Atlantic weather influences remain conducive to the Carignan.

The light soils of the Oligocene-Miocene terrain on the coastal rim, between Sigean and Narbonne, produce easy drinking, less powerful Corbières. The Quaternary terrain of the Lézignan region is noted for its warm, robust wines.

Fitou: A Coastal Sector and a Central Zone

The Fitou AOC covers nine communes that also make the *Vins Doux Naturels*: Treilles, Caves, Fitou, Lapalme, and Leucate in the coastal sector; and Tuchan, Paziols, Villeneuve, and Cascatel in the central zone. While this is a smaller area than Corbières, the soils remain extremely varied thanks to the many different geological formations. Inland, these lean, stony and often shallow soils are derived from schists and limestone on Paleozoic bedrock, plus Triassic marls and hard Upper Cretaceous limestone. The soils of the coastal sector developed from Triassic marls, Urgonian limestone, and Quaternary gravel terraces. The Carignan, blended with the Grenache Noir, flourishes in Fitou. The wines from inland are fine and elegant; those from the coastal sector are powerful and robust.

The castle of Aguilar, perched in the middle of an Urgonian limestone corrie in Tuchan, Fitou AOC.

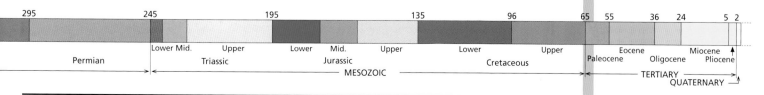

295		245		195		135		96		65	55	36	24	5	2
Permian		Lower Mid.	Upper	Lower	Mid.	Upper		Lower	Upper	Paleocene	Eocene	Oligocene	Miocene	Pliocene	
			Triassic		Jurassic				Cretaceous						
					MESOZOIC						TERTIARY				
											QUATERNARY				

The Minervois and St.-Jean-de-Minervois

The Minervois, that takes its name from the village of Minerve, shares exactly the same geological history as the northern Limousin and especially Cabardès, of which it is the natural eastward extension. While the formations that support the vineyards are very similar, their distribution is quite distinct.

The Minervois resembles a huge stairway running down from the Montagne Noire to the banks of the Aude. The geology once again consists of Thanetian marls and limestone (first marine invasion), especially in the western sector; also Sparnacian red clays and sands, and hard foraminiferal* limestone. This directly overlays the Paleozoic basement in the eastern sector (Minerve, St.-Jean-de-Minervois) since the second marine invasion reached much farther than the first.

The Montagne Noire Lineage

After the seas retreated, the erosional products shed by the growing Pyrenees started to reach the region. Lakes meanwhile formed, which left the lacustrine limestone in Ventenac and the limestones and marls in Agel. Then, the Montagne Noire was uplifted by the aftershocks of the Pyrenean upthrust.

The famous village of Minerve perched on a cliff of foraminiferal limestone at the junction of the gorges of the rivers Brian (right) and Cesse (left). The vines are planted in limestone on the valley sides. Upstream, the Brian flows over the basement rock.

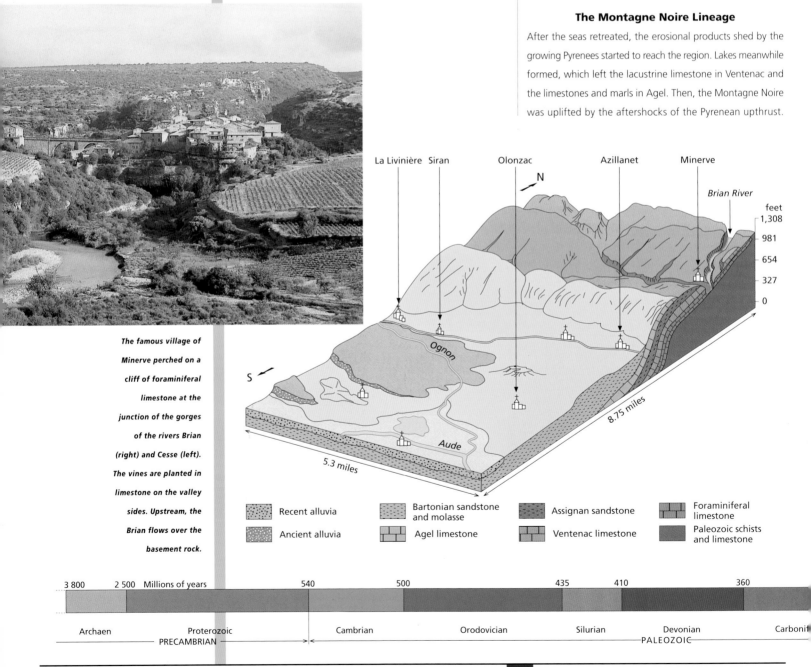

Recent alluvia	Bartonian sandstone and molasse	Assignan sandstone	Foraminiferal limestone
Ancient alluvia	Agel limestone	Ventenac limestone	Paleozoic schists and limestone

3 800	2 500	Millions of years	540	500	435	410	360
Archaen	Proterozoic		Cambrian	Orodovician	Silurian	Devonian	Carbonif
	PRECAMBRIAN					PALEOZOIC	

Along its border, all the layers mentioned above were tilted strongly southward, forming an immense scarp, which stretches from Villeneuve-Minervois to Azillanet and beyond. Rivers eroded these formations, producing deep, steep-sided valleys. A prime example of their handiwork is the spectacular site of the village of Minerve itself. Perched on a cliff of foraminiferal* limestone, the ancient village stands at the junction of the gorges formed by the rivers Cesse and Brian that flow over the basement rock. In the Upper Eocene (Bartonian), the region filled up with molasse that in the Quaternary would be overlaid by stony alluvial deposits. Vast sheets of these alluvia on several levels were spread by the rivers from the Montagne Noire (Glamous, Argent-Double, Ognon, Cesse).

The Three Terroirs of the Minervois

The grand south-facing stairway of the Minervois is interrupted by two north-south landforms that divide the region into three sectors: a western, oceanic sector; a central, more continental sector (dry and warm); and an eastern, Mediterranean sector. The landforms in question are the hills of Villeneuve-Minervois in Trèbes, remnants of a vast, underwater delta formed by a river that flowed down from the Montagne Noire in the Eocene; and the mountain of Serre d'Oupia, folded at the same time as the Alaric (*cf.* Corbières pp. 114-115).

The Minervois then, comprises a relatively limited series of geological formations. The most significant one consists of alternating layers of molasse and banks of Bartonian sandstone. This tougher sandstone layer forms hillocks known locally as *mourrels*, often with a wooded cap of Alep pines, surrounded by shallow, dry soils that particularly suit the Carignan and the Grenache varieties.

Provided yields are kept down, the Quaternary terraces make ideal locations for the Syrah, although the best sites are plainly the piedmont slopes that extend from Caunes-Minervois to Azillanet, and beyond the Serre d'Oupia mountain. This is the sub-regional appellation of La Livinière, producing powerful, deeply colored wines from a blend of the Syrah, planted in

Agel limestone and Assignan sandstone, and the Grenache Noir. Spicy, with an intensely fruity bouquet, these wines take at least two to three years to mature.

St.-Jean-de-Minervois: Lacustrine Ventenac Limestone

In the northeast of the region, the vineyards of St.-Jean-de-Minervois are as unique as they are exceptional. The Ilerdian foraminiferal limestone comes to an abrupt end toward the north, forming dazzlingly white cliffs that stand out vividly against the green of the maquis cladding the dark, Paleozoic Ordovician schists. You get a good view of this contrast from the little road running from St.-Jean-de-Minervois to Belle-Rase. The very hard limestone here is only sparsely planted with vines. Farther south however, limestone overlain by Ventenac lacustrine limestone provides the Muscat à Petits Grains with an exceptional growing environment. This unique geological formation is the source of the most elegant of the Muscat-based *Vins Doux Naturels*, even though the cooler temperature at this altitude represents the limits of this cultivar's cold tolerance. The vines grow exclusively amongst the stones covering the limestone formation, surrounded by a mysterious world of whiteness: white soils, white dry-stone walls surrounding the plots of vines, white house fronts built of hard limestone.

St.-Jean-de-Minervois: the cliffs of Ilerdian foraminiferal limestone rest directly on dark Ordovician schists (valley bottom).

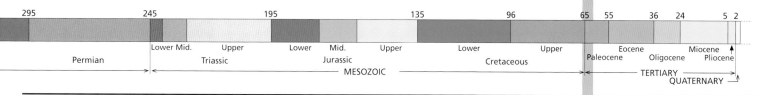

The Carignan

Despite its bad reputation, the Carignan is the most widely planted wine grape in the world, beating the Ugni Blanc. In fact, the Carignan accounts for some 210,000 hectares of planting acreage and exceeded 250,000 hectares before the uprooting policy came into force in the Languedoc-Roussillon. Rarely has a grape variety been more unjustly maligned. Older Carignan vines, planted in carefully selected locations, with limited crop loads and proper wine-making techniques, are capable of producing quality wines.

Originally a Spanish cultivar from Aragon, the Carignan is most widely planted in the French Languedoc-Roussillon region, where it accounts for half of all world plantings and has enjoyed the greatest growth. It is a tough, hardy variety, equally at home on dry slopes or fertile plains, although vines in richer soils yield commonplace wines. The Carignan is really quite a sensitive grape and suffers more than any other cultivar from rough handling when delivered to the collection bays after picking. It produces some of its finest results when vinified by carbonic maceration, a process that involves placing whole, uncrushed grapes into vats filled with carbon dioxide.

In Search of the Ideal Climate

The Languedoc represents the northern limit of cultivation for the weather-sensitive Carignan that prefers not to leave the Mediterranean rim. At the maturation stage, it is especially sensitive to wide fluctuations in day and night temperatures. The Alpilles Mountains, on the border between the Costières de Nîmes and the Rhône Valley, mark the limit of the Carignan's northward development (50-60 per cent in the Costières de

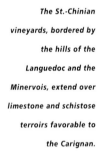

The St.-Chinian vineyards, bordered by the hills of the Languedoc and the Minervois, extend over limestone and schistose terroirs favorable to the Carignan.

The "Crest" de Rivesaltes terroirs around the Salses fortress in Roussillon.

Nîmes AOC compared with just 10 per cent in the Côtes-du-Rhône AOC). What's more, it is not suitable for cultivation throughout the Mediterranean region. A major tasting of more than 250 Carignan-based wines from vineyards between Nice and Perpignan found that quality improved the closer the vine got to the Spanish border. On the basis of this survey, conducted in 1975 when the grand AOVDQS Mediterranean vineyards acquired AOC status, the INAO set approved acreage percentages for Carignan plantings: 30 per cent in the Côtes-de-Provence; 50-60 per cent from the Costières-de-Nîmes to the Minervois; 60 per cent in Corbières; and 70 per cent in the Côtes-du-Rousillon.

Response to Different Terroirs

In the early 1980s, the Belesta Wine Cooperative carried out a comparative study of identically vinified wines from granitic sands or schists, and confirmed the Carignan's super-responsiveness to its terroir. Carignan wines from granitic sands are powerful and tannic and need time to open up, whereas those from schists are friendlier and more appealing. Since then, the Cooperative has successfully marketed a range of cuvées from each soil type. Carignan plantings have declined sharply in recent years, leaving the vineyards with a significant stock of relatively old vines capable of producing powerful wines. Everyone agrees however that the most expressive Carignan wines come from the Côtes-du-Roussillon and Fitou terroirs.

Carignan wines from argillaceous-calcareous soils – Fitou, Tautavel in the Côtes-du-Roussillon-Villages appellation, and the limestone sector of St.-Chinian – have a characteristic fragrance of red berries in youth despite their otherwise subtle nose. With age, they develop aromas of lush, ripe fruits and prunes, overlain with toasty scents of roasted almonds and even leather. The untamed palate of the young wines likewise acquires full and fleshy structure thanks to well-rounded tannins.

Carignan wines from the schistose terroirs of La-Tour-de-France (Côtes-du-Roussillon-Villages) and Tuchan (Fitou) open up more quickly, especially those vinified by carbonic maceration. Notes of pepper and spice combine with elegant, velvety tannins to produce an irresistible wine that will impress even the most hardened Carignan critic. The Faugères wines, also from Carignan planted in schists, have characteristic aromas of burnt pine resin plus solid body thanks to well-rounded tannins. The wines from the Quaternary terraces are robust and for laying down.

The Carignan is now enjoying a revival in the Languedoc after falling out of favor due to recent past difficulties that led many winegrowers to opt for the more beguiling Syrah. The Fitou AOC was the first to make the Carignan an authorized grape variety, followed by growers in Boutenac in the Corbières appellation.

The Roussillon

The Roussillon vineyards cover a 35,000-hectare amphitheater framed by the Corbières to the north, the Canigou to the west and the Albères to the south on the frontier with Spain. The terraced, hilly landscape is the work of three rivers, the Têt, Tech, and Agly, plus other processes that have left a variety of formations.

The Roussillon extends along the axial zone of the Pyrenees, where the mountains reach their highest altitude, and along the *northern Pyrenean zone*. By comparison, the Corbières and the Limousin are exclusively located in the sub-Pyrenean zone. The north Pyrenean zone, that corresponds to the northern half of the Pyrénées-Orientales department, is framed by two major east-west faults that stand out in the landscape. The northern fault roughly follows the departmental boundary line with the Aude. The southern fault runs along the Têt River, from Perpignan to Millas, headed for Bélesta, Montalba-le-Château, and Sournia.

Compartmentalized Topography

The *northern Pyrenean zone* was once a marine trench that gradually became filled with beds of limestone, followed by black schistose marls (foundation of the Maury terroir). This zone, caught in a vise between the axial zone and the sub-Pyrenean zone, was violently compressed and folded in the Pyrenean uplift. Upthrust blocks of basement rock burst through the folded sedimentary covering, producing the Agly massif.

In the south of the department, as the *axial zone* uplifted, it fractured into several blocks separated by east-west faults. The Canigou block was the first to collapse, opening up a vast

The small Maury area at the northeast tip of the Roussillon, one of the region's five Vins Doux Naturels appellations, is predominantly planted with Grenache Noir.

The remarkable site of the Maury vineyards.

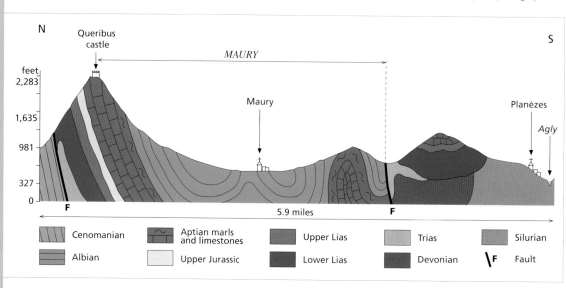

trough in the Roussillon between Perpignan and Le Boulou. South of the fault series, the Albères block, at the frontier with Spain, remained prominent before plunging into the Mediterranean Sea some 18 miles farther east. Now it provides an ideal location for the Banyul vineyards.

At the end of the Tertiary (Pliocene), the Roussillon trough filled up with sediments more than 6,500 feet thick. Marine deposits came first, followed by lacustrine material, then erosional products shed by Tertiary and Precambrian terrain in the surrounding massifs. These deposits were spread by torrents and dumped in a vast sheet that sloped from the piedmont of the Massif du Canigou to the sea. In the Villafranchian (that marks the limit between the Tertiary and the Quaternary) the surface of this area was thoroughly leveled, leaving only a few molasse hillocks (Trouillas, Tresserre, Banyuls-dels-Aspres). Rivers then spread vast sheets of stony alluvia along the length of their banks.

The Schists of Banyuls and Maury

The Roussillon is predominantly known for its Vins Doux Naturels although the Côtes-du-Roussillon-Villages appellation also produces some excellent dry white wines. Banyuls, at the very tip of the Albères block in the axial Pyrenean zone, unquestionably boasts the finest terroir, exclusively supported by lightly metamorphosed Cambrian schists. Thanks to the more or less vertical arrangement of the layers, the roots of the Grenache Noir can penetrate deep down in search of precious moisture. The ingenious schist drainage canals (*peo de gall*) that were built by the Knights Templar in the twelfth century are a unique feature of these steep slopes at the mercy of erosion.

The Maury terroir is also schistose, although composed of Albian marly schists that were folded wallet-fashion between Urgonian limestone and Aptian marls in the course of the violent folding associated with the Pyrenean uplift. What you see today is a spectacular viticultural landscape, where white limestone cliffs – that merge with the fabric of the Queribus castle farther north – frame dark hills of schist, usually with

vertically arranged layers that favor good root development. On both terroirs, the Grenache Noir yields aromatic wines with aromas of prunes, figs, and dried or candied fruit. Banyuls brings out roasted, toasty notes while red berries dominate the Maury wines.

The Côtes-du-Roussillon Mosaic

Looking elsewhere in the Roussillon, we see a different picture to the north and south of the Têt River. The vineyards to the south are mainly established in Pliocene molassic formations, Quaternary terraces, and alluvial cones, with hardly any plantings in the Silurian schists and Devonian limestone of the Massif du Canigou piedmont. The vineyards to the north of the Têt extend over a flat expanse of Pliocene and especially Quaternary terrain known as the "crests" of Rivesaltes and Salses, a vast stony plain of meager soils crossed by the Route Nationale and the freeway to the north of Perpignan. There are more developed vineyards in the mountainous Massif de l'Agly, formed from a block of Paleozoic basement that burst through the Mesozoic covering visible along the mountain rim. It consists of gneiss, granites, mica schist, and schist, and supports the best vineyard plots of the Côtes-du-Roussillon-Villages. The wines from the argillaceous-calcareous soils are powerful and generous while those from schistose soils are more refined and elegant, dominated by aromas of plums and ripe and dried fruits.

Collioure, one of the four villages in the Banyuls appellation and a focal point for many famous painters, has given its name to an AOC that produces dry red and rosé wines from a terroir identical to that in Banyuls. These dizzy slopes of Cambrian schists, carved into terraces by human hand, plunge into the blue waters of the Mediterranean.

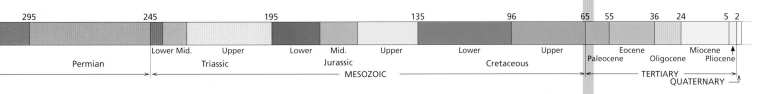

The Muscat Blanc à Petits Grains

The Muscat Blanc à Petits Grains is the most celebrated of the aromatic family of Muscat grapes due to the finesse of the *Vins Doux Naturels* it yields in France. Cultivated in the Mediterranean rim since ancient times, it is still grown in Greece and southern Italy for the production of dessert wines but has also spread to more northerly vineyards where it produces dry white wines.

Only the *Vins Doux Naturels* reveal the full potential of this sweet and fragrant grapevine. Production is concentrated in the Pyrénées-Orientales, the southeast of the Aude and in the departments of the Hérault and the Vaucluse. The leading producer is Muscat-de-Rivesaltes although, uniquely for a *Vins Doux Naturels* appellation, the Muscat Blanc à Petits Grains is not the only authorized grape variety but grows alongside the Muscat d'Alexandrie. The appellation covers a particularly broad area, with vineyards located in the Pyrénées-Orientales and nine communes in the Aude and Fitou region.

A Well Adapted Grapevine

The Muscat Blanc à Petits Grains is a very precocious cultivar that is at ease in almost any location, ripening as easily in the heart of the Roussillon as at the northern tip of the appellation (Fitou region). What's more, it is equally happy in the cool, deep Pliocene molasse of the Aspres region (south of the Têt) as it is in the stony soils of the Rivesaltes-Baixas region northwest of Perpignan. The Tresserre viticultural research center demonstrated that grapes from stony soils contained the highest concentrations of terpenes (geraniol, neroli, and linalol) which are the substances responsible for the characteristic Muscat aromas. The Muscat Blanc à Petits Grains does present one drawback, however, which is that crop yields are very low in poor soils. Growers in the Roussillon prefer the Muscat d'Alexandrie, although conditions here represent the northern limit of the grape's cold tolerance and its performance varies according to location. In the warmest terroirs, it can rival the Muscat Blanc à Petits Grains for terpene concentrations, whereas in deep soils or outlying cooler zones it ripens less easily and crop yields are too high.

A Tradition of Sweet Wines

Outside the Roussillon, only six areas have successfully maintained a tradition of *Vins Doux Naturels* production based on Muscat Blanc: four in the Hérault (Lunel, Mireval, St.-Jean-de-Minervois, and Frontignan), one in the Vaucluse (Beaumes-de-Venise) and one in the Cap Corse region.

The very stony, dazzling white soils of St.-Jean-de-Minervois, derived from weathered Calcaire de Ventenac. The Muscat Blanc à Petits Grains, here planted at the limit of its cultivation area, yields exceptionally elegant, delicate wines.

Strangely, the Muscat Blanc à Petits Grains clings to some astonishing little niches that produce exceptional wines.

The most notable is the Muscat-de-Frontignan terroir, on the southern flanks of the Massif de la Gardiole. Here vines are planted in red soils packed with Jurassic limestone fragments (Oxfordian and Kimmeridgian) or in ancient alluvia, where they yield powerful, highly aromatic VDNs with warm nuances of honey and figs.

The terroirs and the types of wine are the same in the neighboring Muscat-de-Mireval appellation. By contrast, the Muscat-de-Lunel appellation east of Montpellier produces tamer, crisper wines from vineyards planted along a strip of the pebbly overthrust nappe (Villafranchian). This is the only terroir of its type to support Muscat vines.

St.-Jean-de-Minervois marks a return to limestone soils and *Calcaire de Ventenac*. This scrap of terroir at high altitude, located at the limit of Muscat Blanc cultivation, produces exceptionally fine wines with lemony aromas sometimes verging on mint.

The elegant, rose-scented Beaumes-de-Venise, from vineyards nestling at the foot of the Dentelles de Montmirail Mountains (Jurassic), is quite similar in character whereas Muscat-du-Cap-Corse is intensely fruity and fresh.

Muscat dessert wines are widely produced elsewhere in the Mediterranean rim: in Greece (Muscat of Samos, Muscat of Cephalonia, and Muscat of Patras, made from Muscat Blanc à Petits Grains), Italy (Muscat de Pantelleria) and Portugal (Muscat de Setúbal based on Muscat d'Alexandrie). They tend to be richer in sugar with honeyed, nutty aromas of figs and dried fruits.

In the Cap Corse (above) the Muscat Blanc à Petits Grains occupies small niches and part of the Patrimonio area.

In Lunel (opposite) it is established on the stony Villafranchian terrace.

The Picture Farther North

Farther north, the Muscat Blanc à Petits Grains is the basis of the sparkling wines of Die. Clairette de Die Tradition contains at least 75 per cent Muscat Blanc à Petits Grains and starts to bubble due to the naturally occurring grape sugars. The deep, stony soils of Diois suit the sweet southern grape that here yields fine, elegant wines low in alcohol (around 7 per cent). In Italy it yields Asti Spumante from the Piedmont region, a wine very similar to Clairette de Die and made by the same method. Both have fresh, fruity aromas emphasized by a sparkling effervescence. In Alsace, the wines of the Alsace-Muscat AOC combine the liveliness and power of the Muscat Blanc à Petits Grains with the elegant roundness of the Muscat Ottonel.

The Coteaux du Languedoc

T he Coteaux du Languedoc AOC was created by the regrouping of fourteen former AVDQS areas distributed throughout the Languedoc wine-growing region. The intervening sub districts were also later attached. It now consists of 68 communes spread across the slopes and garrigue of a region that extends from the southern border of the Massif Central to the shores of the Mediterranean.

The eleven denominations in the Coteaux-du-Languedoc are: La Clape and Quatourze in the Aude department; and Cabrières, Montpeyroux, St.-Saturnin, Pic-St.-Loup, St.-Georges-d'Orques, Coteaux de la Méjanelle, St.-Drézéry, St.-Christol, and Coteaux-de-Vérargues in the Hérault department. They boast some

remarkable terroirs, mainly producing reds and rosés, with the exception of La Clape and Picpoul-de-Pinet that produce white wines. St.-Chinian and Faugères, in the north of the Hérault and still within the Coteaux-du-Languedoc area, are both appellations in their own right.

A Geological Puzzle

The Coteaux-du-Languedoc, like the Corbières, consists of a real jumble of geological formations, related to a history of turbulence that involved the same major processes already described for the western Languedoc.

• Uplift of the southern edge of the Massif Central dominating the Lower Languedoc region.

• Thrust faulting and overthrust, coinciding with the Pyrenean uplift, which ricocheted as far as St.-Chinian, Gabian, and

The vineyards of the Pic -St.-Loup area, one of the rising stars of the Languedoc. Pictured in the background, the Pic -St.-Loup, on the left, faces Mount Hortus.

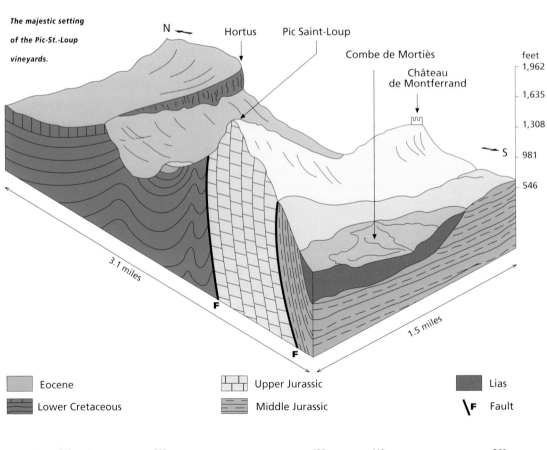

The majestic setting of the Pic-St.-Loup vineyards.

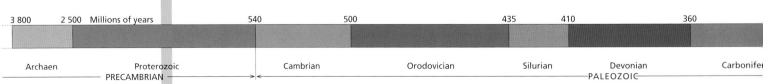

*Faugères, in the Béziers
hinterland, extends
over Visean schists in
the foothills of the
Montagne Noire.*

Montpellier, leaving complex folds of Jurassic and Cretaceous strata. These strata were then laid bare as the folds were dissected by erosion.

There are however differences in regional geology due to the following:

• More significant Jurassic and especially Cretaceous sedimentation, leaving hard limestone deposits characteristic of all the garrigue-covered plateaus to the northeast of Montpellier and in the Gard.

• Oligocene grabens (the Alès and especially the Sommières troughs).

• The collapse of the Pyrenean-Provençal range that disappeared when the Gulf of Lion opened up, this being the most important feature of regional tectonics.

Additionally, practically the entire region was invaded by the seas in the Miocene, with a more timid incursion in the Pliocene. The Quaternary was associated with a more general process of gravelly terrace sedimentation along water courses, especially the Rhône River, that flowed into the sea via Montpellier before adopting its present course.

The Schists of Faugères and St-Chinian

The famous Faugères AOC lies along the southern border of the Montagne Noire, on Paleozoic Ordovician and especially Visean schists (soils rich in gray, yellow and ochre shale platelets). Old Carignan vines blended with the Grenache Noir and Syrah produce highly colored, powerful wines with aromas of the garrigue and red berries. In the east, this terroir extends toward Cabrières (Coteaux-du-Languedoc), famous for rosés dominated by the Cinsault while, in the west, it continues into the northern part of St.-Chinian.

Calcaire Landforms

The southern sector of the St.-Chinian AOC corresponds to the limestone topography of the leading edge of the thrust sheet. The vineyards are planted in argillaceous-limestone soils in small, marly valleys divided by rocky, violently uplifted ridges that face northeast, southwest. The vineyards of St.-Saturnin and Montpeyroux (Coteaux-du-Languedoc) occupy a more

sheltered location away from the edge of the Massif Central, in the foothills of the limestone Grands Causses. There, a mixture of *éboulis* (rubble-like slump material) from the surrounding hard limestone cliffs plus underlying marls, provides the vines with an excellent growing medium. Other parts of the Coteaux-du-Languedoc are less hilly although the soils remain much the same. The soils are very similar in the other, less hilly areas of the Coteaux-du-Languedoc. Good examples are the Cretaceous limestone Massif de La Clape, near Narbonne, and the Jurassic hills of La Mourre and La Gardiole west of Montpellier where the Muscat Blanc à Petits Grains shows a particular liking for the stony soils of Frontignan and Mireval, facing the sea.

This type of *éboulis* is also a feature of the Pic-St.-Loup vineyards (Coteaux-du-Languedoc), that produce truly outstanding wines from stony slopes supplied by the Cretaceous and Jurassic limestone cliffs of the Petits Causses. The peak that gave its name to the region corresponds to the final push of the thrust nappe.

Stony Terraces

The vineyards of the Vérargues region are essentially supported by Oligocene conglomerates composed of rock fragments from the edges of the grabens that opened up in that epoch. Many of the Coteaux-du-Languedoc vineyards however are also established in pebbly, Quaternary alluvial terraces. Good examples are Quatourze in the commune of Narbonne, on an ancient terrace of an abandoned bed of the river Aude; St.-Georges-d'Orques, west of Montpellier; the vineyards on the right-bank of the river Hérault; La Méjanelle, the Montpellier extension of the Costière de Nîmes appellation; and the Lunel region, the only non-limestone terroir planted with the Muscat Blanc à Petits Grains. With age, these wines evolve from a spicy, peppery bouquet of red berries toward scents of leather, nuts and roasted almonds. Wines from limestone soils are closed in their early years, gradually opening to reveal good structure and plenty of strength. Wines from schistose soils tend to be more mineral in aroma with finer tannins, while those originating from *galets roulés* (rolled pebbles) are more forceful in character.

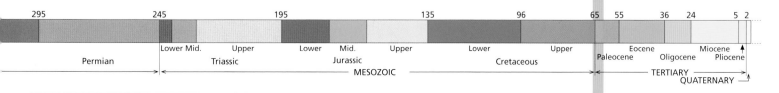

| 295 | | 245 | | 195 | | | 135 | | 96 | | 65 | 55 | | 36 | 24 | | 5 2 |

Permian		Lower	Mid.	Upper	Lower	Mid.	Upper	Lower	Upper	Paleocene	Eocene	Oligocene	Miocene Pliocene
			Triassic			Jurassic		Cretaceous					
						MESOZOIC					TERTIARY		
												QUATERNARY	

Provence

Few wine-growing regions are as varied as Provence with its 43,500 hectares of vineyards covering a patchwork of terroirs between the limestone ranges of Basse Provence – featuring the famous Montagne de St.-Victoire – and the ancient crystalline rocks of the Maures, Estérel, and Mercantour massifs. The range of grape varieties is equally impressive.

The Pyrenean-Provençal range forms part of the Hercynian ranges that came under intense erosion at the end of the Paleozoic Era. Today, the Pyrenean-Provençal range mainly surfaces in the Maures and Estérel massifs, and south of La Seyne-sur-Mer. In the course of weathering under the warm, alternating dry and humid conditions, the range shed sediments that accumulated in dense deposits more than 6,500 feet deep at its northern foot. Gradually, these turned into pale pink sandstone and pelites*. This zone was subjected to greater weathering than the surrounding terrain and today forms the Permian depression that encircles the Massif des Maures from the north.

Major Movements of the Provençal Range

In the Mesozoic Era, Provence was a platform under shallow seas, situated between the much deeper Alpine Sea and the Provençal range that was above water. It accumulated thick limestone deposits.

The Eocene witnessed the major upthrust movements of the Provençal range, with a northward tilting of all the sedimentary strata covering the Paleozoic basement, from the Triassic gysiferous clays upward. As in the Languedoc region, these strata were folded, shattered, and dumped in a heap at the foot of the new range. The forces of erosion then set

The Montagne St.-Victoire, so dear to Cézanne, watches over the pebbly soils of the vineyards that produce excellent reds and rosés.

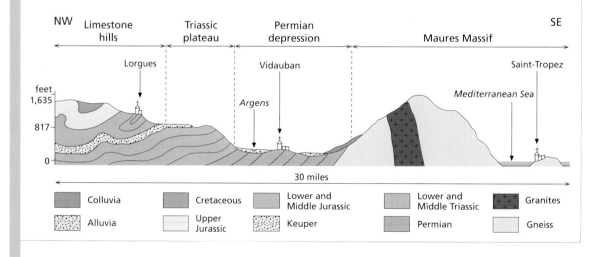

The range of terroirs in Provence.

to work, carving out the sharp outlines of the massifs of St.-Victoire, St.-Beaume, and L'Etoile, around Aix-en-Provence and Marseilles.

The seas would return to part of Provence in the Miocene but the last major events took place at the end of the Tertiary Period. In the south, the Pyrenean-Provençal range collapsed as the Gulf of Lion opened up. All that remains of it, at much lower altitude, are the Maures and Estérel massifs. In the north, the Alps began to uplift. Provence tilted southward, the previously northward-flowing watercourse changed direction and started to flow toward the Mediterranean, and erosion and Quaternary alluvial deposition occurred here as elsewhere.

Five Geological Zones

The highly varied formations that make up Provence today may be divided into five zones.

• The *Massif des Maures*, a homogeneous crystalline complex (granites, gneiss, mica-schist) wrapped in phyllites. The vineyards are mainly established on colluvia, rarely on the poorly developed soils given over to the forest.

• The *Permian Depression* that surrounds the Massif des Maures to the west, north and northeast, from Toulon to Fréjus-St.-Raphaël. The red wine-colored, argillaceous-sandy soils are derived from the red sandstone bedrock (Permian). The colluvial soils, which are the terrain of choice for the production of rosé wines, are derived from the landforms dominating the depression on either side.

• The *Triassic Plateau* and the *Hills of Limestone Provence*, to the north and northeast of the Permian depression. This zone was formed by the violent folding of the Triassic, Jurassic, and Cretaceous limestone into a series of west- and east-facing hills. The different limestone faces are arranged in narrow bands of variable tightness, often with flat-bottomed depressions due to Karstic effects. The landscape is extremely picturesque, with rows of vines on narrow man-made terraces (*restanques**) in small isolated basins owned by particular villages. The vines grow in argillaceous soils derived from angular, decalcified limestone detritus, together with colluvia

and alluvia. This zone also extends into the Coteaux-Varois AOC in the hinterland where the climate is cooler, whereas the previous two zones are exclusively located in the Côtes-de-Provence AOC.

• The *Upper Basin of the Arc* in the Côtes-de-Provence AOC, bordered to the north by the St.-Victoire range and to the south by the Olympe and Aurélien massifs. This is an area of poorly developed soils, and brown soils formed from Upper Cretaceous sandstone and sandstone clays, plus colluvia from the surrounding massifs. It produces exceptionally powerful, elegant wines blended from the Grenache, Syrah, and Mourvèdre. This zone extends beyond Aix into the Coteaux-d'Aix-en-Provence and the Baux-de-Provence, also formed from east-west ranges separated by small, sedimentary basins.

• The *Bassin du Beausset* (Côtes-de-Provence) covers an Upper Cretaceous syncline* that opens to the sea, flanked by the AOCs of Cassis to the west and Bandol to the east. Cassis produces elegant white wines from the Ugni Blanc, Marsanne, and Clairette, planted in Upper Cretaceous marls and sandstone in *restanques* terraces. Bandol on the other hand is the preferred terrain of the Mourvèdre that thrives on limestone slopes of mixed Triassic, Jurassic, and Cretaceous origin.

Additionally, the small Palette appellation in the suburbs of Aix-en-Provence is distinguished by a particular type of geological formation known as *Calcaire de Langesses*. Palette produces robust, fleshy reds, rich, rounded rosés and especially, remarkably harmonious whites.

The vineyards of the Cassis AOC are planted in terraces (restanques) in a corrie of Cretaceous limestone. Cassis white wines are blended from the Clairette, Marsanne and Ugni Blanc and make the ideal accompaniment to a dish of bouillabaisse or seafood eaten al fresco in the port.

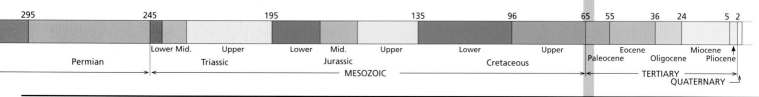

295		245		195		135		96		65	55		36	24		5	2

Permian	Lower Mid. Upper	Lower Mid. Upper	Lower Upper		Eocene	Miocene
	Triassic	Jurassic	Cretaceous	Paleocene	Oligocene	Pliocene

MESOZOIC

TERTIARY

QUATERNARY

The Mourvèdre

The Mourvèdre variety was once widely planted throughout the south of France but teetered on the brink of extinction following the phylloxera epidemic. Today, thanks to the determined efforts of men like Lucien Peyraud, it is the premier grape variety in the Bandol AOC and enjoying a revival throughout the vineyards of the French Mediterranean rim.

The restanques of Bandol, bathed in maritime humidity, where for the past 50 years the Mourvèdre has been regaining its respectability.

Of Spanish origin, the Mourvèdre has a long history of cultivation in the varied soils of Provence, where it was first introduced in the fourteenth century. It yields wines with sustained color and solid tannin texture, two qualities that promise good aging potential.

The Bandol Setting

The Bandol site is ideal for the Mourvèdre. The vines occupy a vast open corrie, shielded from the cold north winds but open to welcome moisture from the sea, so providing the grapes with optimum ripening conditions. There is a rich complexity of soils due to the rugged, craggy geology left by the process of uplift and collapse that caused the Pyrenean-Provençal axis to sink: clays, Triassic cargneules*, Jurassic dolomitic* marls and rudist*-bearing limestone and Cretaceous sandy or argillaceous marls. The compact limestones of the Jurassic and Lower Cretaceous (Urgonian) tend to form rocky ledges more or less crowned with pines and are too dry for the Mourvèdre. It much prefers the deep colluvia at the mid-slope point, and the marlier, almost invariably limestone soils that provide it with the cool, humid conditions it requires to develop good balance. The Mourvèdre has been given a second chance in the coastal vineyards of Bandol, demonstrating yet again that great terroirs are born of a skillful combination of geology, climate, and human talents. The port of Bandol, a natural asset in terms of marketing and wine distribution, was the origin of a special technique designed to speed up the otherwise prolonged aging required by the Mourvèdre. The wines, marked with the initial "B" for Bandol, were placed at the bottom of the ship's hold and sent off around the tropics. When they returned several months later, weather conditions during the passage had accelerated the aging process.

An "Improver" Grape Variety

Since the beginning of the 1980s, the Mouvèdre has attracted renewed interest from winegrowers throughout the Mediterranean rim. Slightly farther inland, there are well-

The vineyards of Jumilla and Yecla, in the Alicante hinterland in Spain, have the largest plantings of Mourvèdre in the world.

developed plantings of the Mourvèdre in the Rhône Valley, notably in Châteauneuf-du-Pape where it has consistently remained one of the premier varieties, thriving in the ideal conditions provided by Villafranchian terraces of deep soils that remain cool in summer. The Mourvèdre, together with the Syrah, belong to a group of so-called "improver" grape varieties that wine-growers in most of the young Languedoc AOCs have been strongly advised, and sometimes obliged, to replant for blending with the Grenache Noir and the Carignan.

Spanish Monastrell

You cannot talk about the Mourvèdre without mentioning Spain, a country where it is more widely planted than anywhere in the world (73,870 hectares compared with 7,300 hectares in France). The Mourvèdre (known locally as the Monastrell) is particularly prevalent in the hinterland of the Alicante region where it accounts for ten times the Mourvèdre acreage in France. There are plantings in the Alicante appellation itself, the Río Vinalopo zone, around Villena and Monovar, and especially in the appellations of Jumilla and Yecla where the Mourvèdre accounts for up to 80 per cent of planting acreage. The arid limestone sierras surrounding these plateaus came under intense erosion in the Pliocene and Quaternary, shedding material that piled up at the foot of the plateaus in a regular slope, rising from 1,144 feet to more than 2,600 feet. The very stony soils in this region feature low-density plantings of ungrafted Mourvèdre vines. The climate is very dry, not to say semi-desert (11 inches of rain per year) and the soils are often raked to conserve moisture. The wines are high in alcohol (13-15 per cent) and were traditionally produced by the *doble pasta* technique: the addition of new vintage wine to a marc of rosé made by the saignée method. The wines were powerful and robust and mainly used to beef up the lightweight La Mancha wines. These days both Jumilla and Yecla areas produce excellent, lower alcohol (12-13 per cent) wines, marketed under their appellation of origin label.

Tuscany and Central Italy

Tuscany, rich in historical and cultural heritage and the symbol of the Italian Renaissance, also boasts one of the finest wine-growing traditions in Europe. Which is not to say that fortune has always smiled on Italian vines. The Chianti is the most widely distributed Italian wine while the Brunello di Montalcino is one of the most representative in terms of quality.

Tuscany is every traveler's dream, with its rolling landscape of soft, verdant hills carpeted with vineyards and cereals, hilltops and avenues of cypress trees. It is located between the Apennine Mountains in the east and the Tyrrhenian Sea – a vast, essentially hilly province consisting of 68 per cent hills, 26 per cent mountains and only eight per cent flat land. It is also a region of infinite variations, with a highly complex geological history linked to the collision of the European and African tectonic plates.

The cellars of Cantina del Redi, beneath the Ricci palace, are dug out of Pliocene sandy sandstone.

The downy slopes of Castello di Ama, in the heart of the Chianti Classico area.

The Apulian Promontory

The Apulian promontory is the name given to the most advanced point of the African plate that punched a hole in southern Europe when the European and African plates collided. Today, it more or less extends into the heel of the famous Italian "boot" and the present-day Adriatic Sea, so occupying the space between Italy and Greece. The promontory underwent subduction, which means that as it advanced northward it sank beneath the European plate, particularly under the eastern part of Italy and the Greek peninsula, from the Peloponnese to Crete and Rhodes.

The Apennine Range

In Italy, this process led to the formation of the Apennine range on the edge of the subduction zone. The uplifting of the Apennine ridge that more or less forms the mountainous backbone of the Italian peninsula is still actively underway, as evidenced by the frequent earthquakes suffered in this area (in the Assise region in 1998, for instance). The uplift and development of the range toward the east coincided with a curious phenomenon that directly affected Tuscany. Just as in the Miocene the Gulf of Lion opened up between the Provence and Languedoc hills, and the Corsica-Sardinia landmass, so in the Oligocene the Tyrrhenian Sea was greatly extended between Corsica and the Apennine range. This formed very deep basins between the Corsican-Sardinian and Italian coasts and also led to the collapse of the whole area behind the Apennines (that continued their eastward

3 800	2 500	Millions of years		540	500		435	410		360

Archaen	Proterozoic	Cambrian	Orodovician	Silurian	Devonian	Carbonif
	PRECAMBRIAN				PALEOZOIC	

Tuscany's gradual collapse into blocks separated by faults has left a landscape of endless slopes watched over by ranks of towering cypress trees.

development). One notable case was the Tuscany landmass which fractured into a series of blocks separated by faults running northwest and southeast parallel to the ridge of the Apennines. The blocks gradually sank, leaving a region with a very particular geological make-up.

Tuscany: Continental and Insular

In this part of Italy, the structure of the Apennine range on the Tuscany side is mainly composed of an Oligocene sandstone formation, also known as *macigno*, (millstone). Most of the mountains are smooth and rounded even though some can reach heights of 6,500 feet or more. They extend toward the northwest into the towering limestone and dolomite peaks of the Apuanian Alps, source of Italy's world-famous Carrara marble. Tuscany, located on the range's western flanks, essentially consists of more or less sandy Tertiary and Quaternary mixed clay formations.

The region also includes seven islands in the Tyrrhenian Sea: Elba (the biggest, with *Denominazione di Origine Controllata* denomination), Giannutri, Giglio, Capraia, Pianosa, Montecristo, and Gorgona. All of these islands vary in geological make-up. Elba is formed from granites, schists,

ophiolites*; Montecristo and Giglio are essentially granitic; Giannutri and Capraia are, respectively, limestone and volcanic in origin.

The Chianti Area

Chianti, Tuscany's largest denomination of origin area, produces more than two million hectoliters of wine a year. There are two characteristic zones: Chianti Classico, corresponding to the original appellation area, and the much more extensive Chianti region, where the soils are somewhat different.

The Chianti Classico area lies around the Chianti Hills, within an area defined by Florence in the north, Arezzo in the east, and Poggibonsi and Sienna in the southwest. It is bordered to the east by the waters of the Arno River that supply Florence, and to the west by the Val d'Elsa, through which run road and rail links from Sienna to Florence. The southern limits are roughly located in the middle of the Tuscany Hills. The Chianti Classico terroirs are mainly composed of Tertiary schistose clay, known locally as *galestro*, that particularly suits the Sangiovese grape. Here it is blended with the Canaiolo, together with French varieties such as the Cabernets

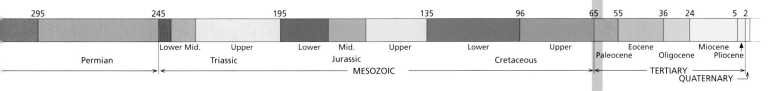

295		245		195		135		96		65	55		36	24		5	2
	Permian	Lower	Mid.	Upper	Lower	Mid.	Upper	Lower		Upper	Paleocene	Eocene	Oligocene		Miocene	Pliocene	
			Triassic			Jurassic			Cretaceous								
						MESOZOIC							TERTIARY				
														QUATERNARY			

The vineyards of Castello Banfi, one of the largest estates in the Brunello di Montalcino appellation.

Vino Nobile di Montepulciano is produced on the valley slopes running from the magnificent hilltop village of Montepulciano toward the plain of Val di Chiana.

The slopes of Montepulciano.

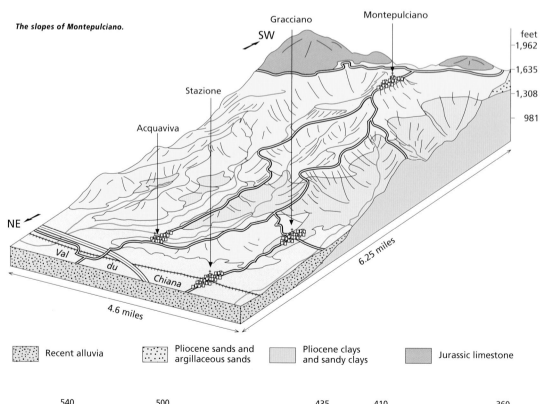

Recent alluvia

Pliocene sands and argillaceous sands

Pliocene clays and sandy clays

Jurassic limestone

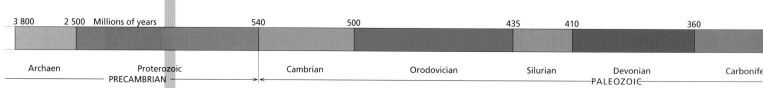

| 3 800 | 2 500 | Millions of years | 540 | 500 | 435 | 410 | 360 |

| Archaen | Proterozoic | Cambrian | Orodovician | Silurian | Devonian | Carbonife |
| PRECAMBRIAN | | | PALEOZOIC | | | |

(Sauvignon and Franc) and the Merlot. The highest peak is Mount Calvo (2,740 feet) that dominates a verdant landscape of vineyards alternating with fields of crops and woods.

Vino Nobile di Montepulciano and Brunello di Montalcino

These two famous DOCGs (*Denominazione di Origine Controllata e Garantita*) lie south of Sienna, where the climate is distinctly drier and even arid. Vino Nobile di Montepulciano, the longest established, is exclusively located in the commune of Montepulciano and in the flat part of the Val di Chiana beyond the railway line. The terroir surrounds the superb village of Montepulciano, perched on a promontory that extends north-south. It is predominantly composed of east-facing slopes of Pliocene sandy and argillaceous-sandy formations that rise 815-1,962 feet. Average annual rainfall is about 27 inches. The Prugnolo Gentile (Sangiovese clone), Canaiolo Nero, and Mammolo yield dark red, powerful, and tannic wines that need time to mature.

Brunello di Montalcino, a few miles farther west as the crow flies, covers a much wider area although still centered on the hilltop village of Montepulciano. The vineyards mainly lie on the south-facing, argillaceous-limestone slopes running down from the village to the River Orcia that marks the appellation's southern border. The Orcia discharges into the River Ombrone that borders the appellation to the west and north. This is a significantly drier region (average annual rainfall of 19 inches), exclusively planted to the Sangiovese, known locally as the Brunello. The wines are powerful and slow to open, in contrast to the less robust wines from the more recently established vineyards in the north and south of the appellation.

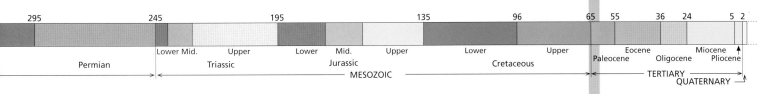

The Rhône Basin

There are fundamental differences between the Rhône Valley and the Paris Basin or the basin of Aquitaine, even though all three are sedimentary basins. Witness the Rhône Valley's more usual name of Rhône Corridor. This is a sedimentary basin with a turbulent history, wedged tight between the young Alpine range to the east and the ancient, virtually unshakable bedrock of the Massif Central to the west. The southern part – south of Valence – differs significantly from the northern part.

Southeastern France is divided in two by a major line of deformation in the Hercynian basement passing through Alès, Privas, La Voulte-sur-Rhône, south of Valence, then beyond the Rhône, through the Isère Valley and Chambéry. In the Jurassic, the entire region was covered by seas. But in the Lower Cretaceous this Hercynian deformation separated two zones of distinct sedimentation, with shallow waters to the north, between Valence and Lyon, but deep seas to the south in the Vocontian Trough, named after the tribes who inhabited Diois, where the trough was at its deepest.

Around the Vocontian Trough

This trough was surrounded on all sides by a continental platform under shallower waters where reef limestone was deposited. These are the origins of the very hard limestones, more than 1,300 feet thick, that now form a belt of impressive, often sheer-sided hills, entirely surrounding the Rhône Valley. In the north, lie Mount Vercors, the Vivarais plateau and the Nîmes garrigues, and in the south, the Alpilles, the Luberon and Mount Ventoux.

In the Upper Cretaceous, the Vocontian Trough filled up with the sandstone and sandy limestone or marly-sandstone formation that we see today in the center of the Urgonian* belt, forming the right-bank of the Rhône, Tricastin and the Massif d'Uchaux. Provence then underwent the first phase of folding caused by the uplift of the Pyrenean-Provençal axis that forced the Jurassic and Cretaceous cover northward in long parallel folds running in an east-west direction (Luberon, Ventoux, and Lure mountains). This phase lasted until the end of the Eocene.

Rhône Corridor Formation

In the Oligocene, thick deposits of conglomerates, sandstone and marls, together with lacustrine limestone, accumulated in the piedmont of young hills in emerged regions, and the Rhône Valley axis collapsed. Part of this depression filled up with erosional products but it essentially provided a ready-made corridor for the sea that returned with a vengeance in the Miocene. The waters invaded a vast area and met up, via

3 800	2 500	Millions of years	540	500	435	410	360

Archaen	Proterozoic	Cambrian	Orodovician	Silurian	Devonian	Carbonife

PRECAMBRIAN   PALEOZOIC

The village of Beaumes-de-Venise nestles at the foot of the Massif des Dentelles de Montmirail. This grandiose setting is the home of the Muscat de Beaumes-de-Venise Vins Doux Naturels, and some excellent red Côtes-de-Rhône-Villages.

The very hard Urgonian limestone belt surrounding the southern Rhône Valley is a feature of the Côtes-du-Vivarais AOC, planted in red, pebbly soils.

present-day Switzerland, with an immense sea that engulfed the whole of central Europe. Sands, marly sands, and sandy molasse were deposited across many hundreds of feet, leaving the sediments that now line the bottom of the region's two principal depressions, the Valréas-Vaison-la-Romaine Basin and the Avignon-Carpentras Basin. These deposits also occur on the borders of the Luberon Massif and to the south of Valence. The Alpine uplift then reached a crescendo. The Mediterranean Sea, now cut off from the Atlantic Ocean, temporarily shrank and the Rhône, that more or less followed its present course, dug itself not only through the Miocene formations but also through the Urgonian limestone (formation of the Donzère defile). The deep narrow gorges then occupied by the Rhône would provide a gateway for the final marine incursion in the Pliocene that deposited fine argillaceous and argillaceous-sandy elements.

Terrace Formation

The Alpine uplift led to the raising of the region along its eastern border, intensifying erosion. Powerful rivers rushed through the Rhône Corridor, washing along large masses of material that we now see in the sheets of quartz pebbles mixed with red, sandy clay that form the Villafranchian terracing.

In the Quaternary, erosion changed the overall look of the landscape. Rubble and angular scree built up in the piedmont of the Urgonian limestone hills, while four levels of stony terrace systems were formed along the Rhône and its tributaries.

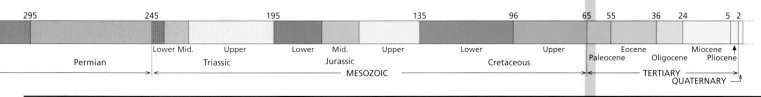

295		245			195				135		96		65	55		36	24		5	2

Permian	Lower	Mid.	Upper	Lower	Mid.	Upper	Lower	Upper	Paleocene	Eocene	Oligocene	Miocene	Pliocene
	Triassic			Jurassic			Cretaceous						

MESOZOIC

TERTIARY

QUATERNARY

The Southern Rhône Valley

The Côtes-du-Rhône appellation, France's Number Two wine-growing area, mainly occupies the southern part of the Rhône Valley that broadens into an embayment south of the Donzère defile, near Montélimar. The climate here is Mediterranean, much to the liking of the Grenache, the region's premier grape variety that thrives in the varied soils.

The southern Rhône Valley is fairly straightforward in geological terms although there are subtle differences between conditions on either side of the river. A massive ring of Lower Cretaceous Urgonian limestones towers over the valley on all sides, rising up like the sides of a huge container.

Urgonian Limestone

These very hard limestones tend to produce skeletal soils that, with the exception of the Bois-St.-Victor plateau to the west of Tavel, support very few of the Côtes-du-Rhône vineyards. The flat Urgonian limestones on the right bank are essentially occupied by the Côtes-du-Vivarais vineyards, where the Grenache Noir and the Syrah do well in the lean, red pebbly soils.

The limestones on the left bank were uplifted, some more steeply than others (Luberon, MountVentoux, Montagne de La Lance). More violent folding even exposed the Jurassic and Triassic strata in the Massif des Dentelles de Montmirail, an ideal rock climbing location. The vines cluster around a vertical slab of Portlandian limestone, shielded from cool alpine weather by their rocky screen.

Upper Cretaceous Valleys and Gravelly Terraces

The Urgonian limestones on the right-bank surround Upper Cretaceous hills. Rivers have carved deep gorges in these limestones (gorges of the Ardèche and the Cèze), opening up wider valleys in the softer, Upper Cretaceous formations and depositing Quaternary terrace zones near the Rhône. The valley slopes are thickly planted with vines that begin on the gravelly Quaternary terraces of the Ardèche, the Cèze, and the Rhône, then scramble up through Upper Cretaceous sandstone and limestone rubble, sometimes mixed with loess. Some go higher still - the Cellettes viticultural area in St.-Gervais, for instance, is placed on the windswept plateau right at the top.

Two Molasse and Sand Basins

On the left bank, the Upper Cretaceous is only found in the Tricastin area (department of the Drôme), and the Massif d'Uchaux where it underpins Mornas castle, perched on the sandstone cliffs that follow the A7 freeway. The scenery along the banks of the Rhône in the Drôme and the Vaucluse is in fact completely different and marked by two huge depressions. One is the Visan-Valréas basin while the other is the flat-bottomed Avignon-Carpentras basin with its lining of essentially sandy Helvetian (Miocene) molasse together with Pliocene sands along waterways. These sediments are the source of light, fruity, and quaffable wines blended from the Grenache Noir and Syrah. Between the rivers Ouvèze and Eygues, and between Eygues and the Valréas basin, we find the vines doing well in stony, colluvial soils on hills formed from Upper Miocene marls and conglomerates that resisted erosion.

Terraces of *Cailloux Roulés*

The area overall is extensively covered with Villafranchian and Quaternary gravel alluvia. The large alluvial deposits at the entrance to both basins were dumped by the Rhône and its tributaries, the Lez, Eygues, and Ouvèze, discharging from gorges of varying depth and narrowness carved out of the surrounding banks of hard limestone. These deposits form vast flats at the bottoms of the basins. There are also several levels of gravelly colluvia (Mindel, Riss, Würm) at the foot of the Massif des Dentelles de Montmirail, from Séguret to Gigondas and Vacqueyras; also in the Plan de Dieu vineyards, at the foot of the hilltop villages of Cairanne and Rasteau;

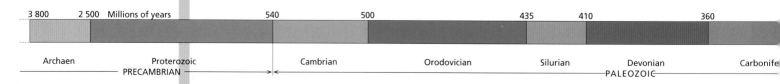

3 800	2 500	Millions of years		540	500		435	410		360

Archaen	Proterozoic	Cambrian	Orodovician	Silurian	Devonian	Carbonife
	PRECAMBRIAN				PALEOZOIC	

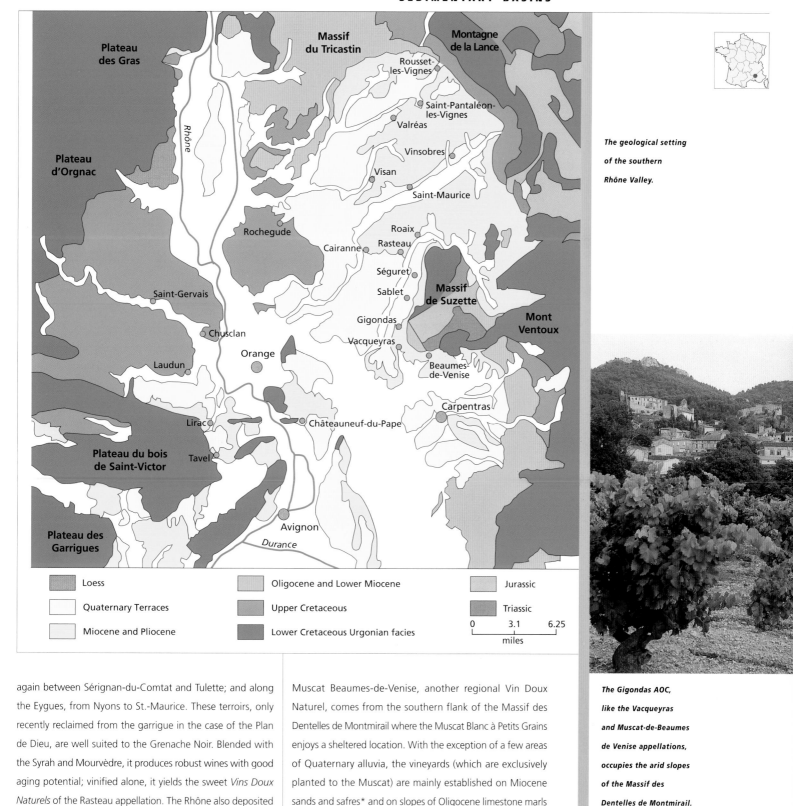

The geological setting of the southern Rhône Valley.

Legend:

- Loess
- Quaternary Terraces
- Miocene and Pliocene
- Oligocene and Lower Miocene
- Upper Cretaceous
- Lower Cretaceous Urgonian facies
- Jurassic
- Triassic

0   3.1   6.25
miles

The Gigondas AOC, like the Vacqueyras and Muscat-de-Beaumes de Venise appellations, occupies the arid slopes of the Massif des Dentelles de Montmirail.

again between Sérignan-du-Comtat and Tulette; and along the Eygues, from Nyons to St.-Maurice. These terroirs, only recently reclaimed from the garrigue in the case of the Plan de Dieu, are well suited to the Grenache Noir. Blended with the Syrah and Mourvèdre, it produces robust wines with good aging potential; vinified alone, it yields the sweet *Vins Doux Naturels* of the Rasteau appellation. The Rhône also deposited large amounts of Quaternary alluvia in Châteauneuf-du Pape, Tavel, and Lirac (*cf.* p166).

Muscat Beaumes-de-Venise, another regional Vin Doux Naturel, comes from the southern flank of the Massif des Dentelles de Montmirail where the Muscat Blanc à Petits Grains enjoys a sheltered location. With the exception of a few areas of Quaternary alluvia, the vineyards (which are exclusively planted to the Muscat) are mainly established on Miocene sands and safres* and on slopes of Oligocene limestone marls and conglomerates – source of the Muscat's characteristically musky, rose-scented bouquet.

295		245		195		135		96		65	55		36	24		5	2

Permian | Lower Mid. / Upper | Triassic | Lower / Mid. / Upper | Jurassic | Lower / Upper | Cretaceous | Paleocene | Eocene | Oligocene | Miocene / Pliocene

MESOZOIC — TERTIARY — QUATERNARY

The Vineyards of the Pre-Alps

The Rhône Valley is bordered on its eastern flank by a series of mountainous massifs that reach a peak with Mount Ventoux, the 6,242-foot "Giant of Provence". Within this mountainous setting are the vineyards of the southern Côtes-du-Rhône, away from the river, and the up-and-coming AOCS of the Côtes-du-Luberon, Côtes-du-Ventoux, and Diois region. The wines from these areas feature unique characteristics that have as much to do with the terroirs as the cooler nature of the climate.

The region lies at the junction of two major geological processes that left their mark on the French landscape: the Pyrenean-Provençal uplift and the Alpine uplift. The first vertical throw toward the north coincided with the uplift of the Pyrenean-Provençal range, now almost entirely engulfed by the Mediterranean Sea (outlining the Luberon and the Ventoux). The second vertical throw toward the south or southwest, coinciding with the Alpine uplift, was increasingly

Clairette-de-Die and Crémant-de-Die both originate from hilly slopes that felt the violent rockings of the Alpine upheaval.

pronounced from the Luberon to the Diois region where the level of disturbance was extreme.

The Luberon

The Jurassic formation in the range of mountains on the right bank of the Durance River, from Mirabeau to Cavaillon, is of no viticultural significance. It outcrops only at the heart of the Pont de Mirabeau defile and within the hollow anticline* at Beaumont-de-Pertuis. The massif extends east-west, mainly supported by a backbone of Cretaceous limestones: Urgonian limestones in the case of the Petit Luberon; Hauterivian limestones, sometimes mixed with marls, in the case of the Grand Luberon. This area is free of vines, except for a small, high-quality vineyard southwest of Apt where the soils are a mixture of sands, flints, and limestone fragments. The bulk of the vineyards in the Calavon valley, both on the southern and northern flanks of the Luberon, rest on two types of soils: on the one hand, sandy soils derived from Miocene sediments, with aspects favoring high-quality viticulture (Aigues region); and on the other, stony soils on ancient terraces located on strips of Pliocene glacis and on alluvial fans in the piedmont. Additionally, the area between Lacoste and Ménerbes includes a few patches of Eocene limestone overlain by red soils mixed with stony fragments. This is a predominantly red wine region although with a surprisingly high proportion of white wines (22 per cent) thanks to the soil and a climate that is cooler here than along the Mediterranean coast. The red wines, produced from the Grenache and Syrah that thrive in these conditions, are full and generous with a solid core of silky tannins and aromas of red berries and spice.

The Côtes du Ventoux

The Massif du Ventoux, like the Massif du Luberon, essentially consists of Urgonian limestones that reach a height of more than 6,500 feet at the crest of the mountain. Viewed from the plain, the Giant of Provence appears to be capped with snow year-round due to the covering of loose stones produced by the frost-shattered, dazzlingly white rock at the summit. Once again, there are rarely any vines in the limestone terrain. Most of the vineyards

3 800	2 500	Millions of years		540		500		435	410		360
Archaen		Proterozoic			Cambrian		Orodovician		Silurian	Devonian	Carbonife
	PRECAMBRIAN									PALEOZOIC	

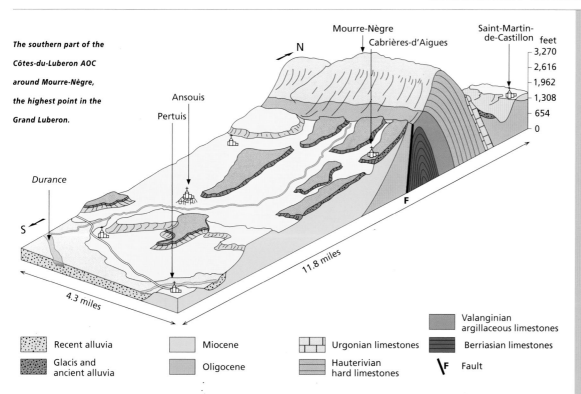

The southern part of the Côtes-du-Luberon AOC around Mourre-Nègre, the highest point in the Grand Luberon.

Legend:
- Recent alluvia
- Glacis and ancient alluvia
- Miocene
- Oligocene
- Urgonian limestones
- Hauterivian hard limestones
- Valanginian argillaceous limestones
- Berriasian limestones
- F — Fault

are planted in the Tertiary sedimentary soils mainly found in the vast crescent shape formed by Mount Ventoux and on the right bank of the Calavon, from Apt to Gordes. The remaining vineyards are located around Carpentras, Mazan, St.-Didier, and Pernes on extensive flats of red, stony soils derived from Quaternary alluvia. As before, production is dominated by red wines. Wines from the AOC Côtes du Ventoux combine spicy, fruity aromas and hints of liquorice and woodiness with quaffability– characteristics that have more in common with the Côtes-du-Rhône wines than those of the Côtes-du-Luberon.

The picturesque village of Gordes in the Côtes-du-Ventoux AOC, stands guard over the entrance to the Calavon Valley, between Ventoux and Luberon.

The Diois

The Diois has two parts: a foreland region featuring the east-west folds of the Provençal movement, especially the spectacular hilltop syncline of the forest of Saou; and the Diois proper, commencing at Saillans, on a juxtaposition of east-west-facing and northwest-southeast-facing folds. This distinction is reflected by the course of the Drôme River that flows east-west down the slope, then north-south from Saillans, making a huge loop at the level of Die. The Diois vineyards are supported by two main terrain types. The first is well suited to the Muscat grape varieties and consists of grayish-blue schistose marls alternating with accumulations of Jurassic and Cretaceous argillaceous limestones at the bottom of the Vocontian basin. The second, as in the Côtes-du-Luberon and Côtes-du-Ventoux appellations, corresponds to erosional products that play a major role in local geology and take the forms already described: fluvial terraces; éboulis (scree, rubble) on limestone ledges mixed with marls from the slopes; and alluvial cones. The most favorable vineyard sites are on south-facing flanks in small Drôme River tributary valleys. The local Clairette produces the light, delicately effervescent Crémant-de-Die with its notes of green fruits. It shares the terroir with the Muscat that excels in the Clairette-de-Die AOC, yielding a rose-scented wine with lemony, citrus aromas and all the flavor sensation of biting into crisp fruit.

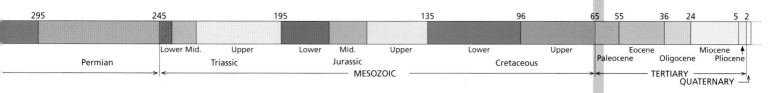

| 295 | 245 | 195 | 135 | 96 | 65 | 55 | 36 | 24 | 5 | 2 |

Permian | Triassic (Lower Mid. Upper) | Jurassic (Lower Mid. Upper) | Cretaceous (Lower Upper) | Paleocene | Eocene | Oligocene | Miocene Pliocene

MESOZOIC — TERTIARY — QUATERNARY

The Vineyards of La Rioja and Navarre

The Ebre Valley is one of Spain's northern-most wine-growing regions, and the vineyards here unquestionably produce the country's most highly regarded red wines. This is where we find the celebrated vineyards of La Rioja, Navarre, and, farther downriver, Campo de Borja, Catalayud, and Cariñena. In terms of geology and pedology, the region has much in common with the southern Rhône Valley.

The Ebre rises at an elevation of some 6,590 feet, in Peña Labra in the Cantabrian Mountains, about 30 miles south of Santander. The river flows rapidly from its source in the Hercynian part of the range toward the Mesozoic limestone ranges (Jurassic and Cretaceous) of the Basque Pyrenees. It crosses the Oligocene depression of Villarcayo (1,962-2,289 feet high), then flows through picturesque gorges into the Miranda de Ebro depression (1,635 feet high). These two depressions, despite a few vines here and there, are not part of the La Rioja vineyards. These only commence a few miles

In La Rioja Alta, vines planted in Miocene marls and sands climb steadily up the slopes (pictured here: Cenicero, on the right bank of the Ebre).

3 800	2 500	Millions of years		540		500		435	410		360	

Archaen	Proterozoic	Cambrian	Orodovician	Silurian	Devonian	Carbonife

PRECAMBRIAN   PALEOZOIC

farther downriver, beyond the limestone gorges of Conchas de Haro carved out of the Montes Obarenes. From there, the Ebre flows into the Logroño basin that opens out toward the east, and eventually into the great basin of Saragossa.

A Basin Framed by Two Ranges

In 1991, La Rioja was the first appellation to receive *Denominacion de Origen Calificada*, Spain's highest wine classification. The province of La Rioja – named after the River Oja, a tributary of the Ebre – is self-governing and forms part of Castilla la Vieja. With vineyards in the south, in La Rioja, and in the north, in the Basque Country and Navarre, the overall wine-growing area is divided into three sub-sectors: La Rioja Alavesa, La Rioja Alta, and La Rioja Baja.

The right bank, upriver from Logroño, is entirely located within the Rioja Alta area (La Rioja Province) whereas the left bank, from Haro to Logroño, is the domain of La Rioja Alavesa plus a small corner of the La Rioja Alta zone. Downstream from Logroño lie the vineyards of La Rioja Baja. These divisions are for administrative purposes only and unrelated to the soils found in these areas or to the wines that are produced there. The vineyards occupy a vast triangular depression lined with Tertiary sediments (Oligocene and Miocene) and squeezed between two mountain ranges, the Cantabrian Mountains to the north and the Iberian range to the south. Both ranges were formed in the course of the Pyrenean uplift by the increasingly violent compression between the African plate and the European plate.

The Iberian range starts in the west with the Sierra de la Demanda that divides Logroño from Burgo and towers over the La Rioja Alta region. The 7,400-foot La Demanda massif is a block of Hercynian bedrock that was leveled and smoothed in the Permian. Formed of Cambrian schists and quartzite, it burst through to the surface in the course of the Pyrenean uplift, ripping apart the overlaying Mesozoic strata. The marked difference in level between the huge fault running along the mountain's northern edge and the Tertiary basin is clearly visible in good weather. The Mesozoic terrain along the fault,

together with some Oligocene conglomerates, were forced toward the vertical by the violence of the uplift. There are no vineyards in either the Paleozoic or Mesozoic terrain or indeed at any height exceeding 1,962 feet.

Farther east, the terrain shows a comparatively smoother topography at the foot of the Sierra de la Cebollera. The massif is formed of Cretaceous limestones bordered by Jurassic terrain

Large areas of La Rioja Alta (such as here, near Najera) are covered with Quaternary alluvia overlaying Miocene sandy marls.

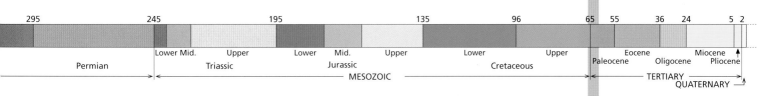

295		245		195		135		96		65	55		36	24		5	2

Permian		Lower Mid.	Upper	Lower	Mid.	Upper	Lower	Upper	Paleocene	Eocene	Oligocene	Miocene Pliocene
		Triassic			Jurassic		Cretaceous					

MESOZOIC — TERTIARY — QUATERNARY

The La Rioja Qualified Denomination of Origin area (Denominación de Origen Calificada) starts a few miles west of Haro, past the narrow gorges of Conchas de Haro (pictured here in the background) that frame the course of the river Ebre.

Facing page: in Rioja Alta (left bank), the vines extend over yellow Miocene sands at the foot of the village of San Vicente de Sonsierra.

that descends much lower where it meets the Oligocene sandstone. The vineyards of La Rioja Baja extend from heights of 1,635 feet in this sandstone formation, down to the Ebre River in the valley below.

The basin on the left-bank is surrounded by various hills. The range crossed by the Ebre on its way to the Conchas de Haro is a small cordillera of Cretaceous (Sénonian) limestone mountains, not exceeding heights of 4,695 feet, called the Sierras de Toloño, de Cantabria and Codes. The southern edge of these limestone layers is sharply sheared by a fault, creating a ledge of small massifs, some craggier than others. They now dominate the Tertiary basin and its vineyards, from Labastida to Laguardia, shielding them from cold north winds.

The vineyards of Navarre are the northward and eastward extensions of the La Rioja vineyards. The Tertiary basin of the Ebre widens substantially here, with the bulk of the vineyards extending north of the river as far as the first Eocene limestone ranges surrounding Pampeluna – where the Oligocene-Miocene basin becomes very wide indeed. The Navarre vineyards also extend south of the river, in the Riberas de Tudela region immediately adjoining La Rioja Baja.

Tertiary and Quaternary Terrain

The terroirs of La Rioja and Navarre, sandwiched between the Iberian chain and the foothills of the Pyrenees, are formed on the Oligocene and Miocene fill of the Ebre depression, crowned by gravelly Quaternary terraces along the rivers. Upriver from Logroño, the climate is temperate, with mean annual temperatures of 55-55.4°F and average annual rainfall of 19 inches. This area produces the finest wines, from vineyards spread throughout the depression. From the end of the Tertiary to the beginning of the Quaternary, the Ebre and its tributaries deposited vast layers of gravelly material in La Rioja Alta and La Rioja Alavesa. Then the rivers cut through these deposits and the soft Tertiary material directly beneath (marls, argillaceous sands, sandstone). Gradually they carved out broad valleys south of the Ebre, narrower ones to the north, while also laying down fresh deposits of terrace fill (rolled

3 800	2 500	Millions of years		540	500		435	410		360
	Archaen	Proterozoic		Cambrian	Orodovician		Silurian	Devonian		Carbonife
		PRECAMBRIAN						PALEOZOIC		

pebbles) on several levels. The remnants hills that resisted erosion now form tabular-shaped landforms with perfectly flat summits that are highly characteristic of the Rioja Alta landscape. The vineyards of La Rioja Alavesa, shielded behind the Sénonian limestone ledge, mainly extend over yellowish, well-exposed Miocene soils, partly covered with rolled pebbles. Planted in Tertiary formations and stony Quaternary terraces alike, the vines reign supreme near the river but increasingly give way to cereal crops at higher altitudes. The Tempranillo is the predominant variety, beating the Mazuelo and the Graciano. The wines from this region are of the highest quality and intended for many years of aging in the wood. They are classified as *Crianza*, *Reserva* or *Gran Reserva* depending on the length of maturation in American oak and in bottles. Bright red La Rioja wines are powerful and elegant, with well developed notes of ripe fruits, cinnamon, cloves, and pepper, overlain by a discreet woodiness.

The climate is warmer and drier in La Rioja Baja where the Tempranillo vies with the Grenache. The landscape changes and, due to the low rainfall, irrigation is essential for crops other than vines (asparagus, artichokes, bell peppers). In these dry conditions, the truck crops grow in green strips alongside the water whereas the vines, mingled with fruit and olive trees here and there, are planted in reddish-gray soils mixed with rolled pebbles on the terraces, and sometimes in sandy, Miocene soils.

The Navarre vineyards are very similar in appearance to those of La Rioja Baja. The climate there ranges from sub-desert in the southeast in Bardenas Reales to cooler and less dry at the foot of the Pyrenees.

Riberas de Tudela, south of the Ebre, like the southern part of Navarre, features river valleys of variable width with villages located along the valley margins. The Grenache reigns supreme once again, essentially established on the steep, stony terraces that mark the rivers Ega, Arga, Zidacos, and Aragón, flowing down from the limestone foothills of the Pyrenees.

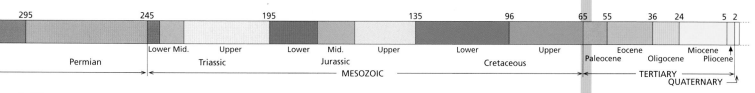

The Italian Piedmont Region

The fairy tale landscape of the Italian Piedmont is located where the Pô Plain meets the foothills of the Apennines. The region produces a range of famous wines, most notably the celebrated Barolo that is now regarded as one of Italy's finest red wines. The Piedmont is also reputed for its Asti Spumante and Moscato d'Asti, produced a few miles farther east, beyond Alba.

The Piedmont is a land of milk and honey. Bra is the center of a major cheese festival, while Alba is famous for its mushrooms that fill shop windows in the autumn. Its white truffles in particular attract tourists from all over Europe. Vines are equally part of the tradition and have been grown in the Piedmont for centuries. In fact, the Lur-Saluces family, who made Sauternes famous, came originally from Saluzzo, just a stone's throw from Alba.

The Pô Valley

The Barolo region, like the rest of the Piedmont, forms part of the upper Pô Valley that extends for hundreds of miles eastward, gradually opening out on its way to the Adriatic Sea. On either side of this immense basin are two mountain ranges that played a major role in its formation. To the north, you have the Italian-Swiss Alps followed by the Italian-Austrian Alps that boast some of the highest peaks in Europe. To the

The vineyards of the Piedmont carpet the Po Plain, set against the majestic backdrop of the Alps.

3 800	2 500	Millions of years	540	500	435	410	360
Archaen		Proterozoic	Cambrian	Orodovician	Silurian	Devonian	Carbonif
	PRECAMBRIAN				PALEOZOIC		

south are the more modest Apennines that barely reach heights of 6,540 feet between Bologna and Pisa, dropping to around 3,270 feet south of the Barolo region. The collapse of the vast Pô depression occurred in two stages that coincided with the formation of these two ranges.

The first stage was associated with the formation of the Alps that exerted an influence in the Oligocene and at the start of the Miocene. Before the climax of the Alpine uplift on the borders of France and Switzerland came a period of crustal stretching that weakened the earth's crust, causing it to collapse behind the Alpine nappes as they advanced toward the west and north. The second stage involved the Apennine range that took up its position behind the Alpine range, this time advancing toward the northeast. The lithosphere (Earth's crust) sagged beneath the accumulated weight of layer upon layer of Apennine nappes, sinking deeper and deeper closer to the range itself – imagine a weak, muddy road surface that is gradually crushed by the advancing wheels of a car. This is why the Tertiary sediments along the range can be more than 32,722 feet thick in places. The collapsing basin gradually became filled with these sediments consisting mainly of molasse from weathering of the surrounding rock. In the Quaternary, the Tertiary formations around the basin were dissected by the Pô and its tributaries.

The Barolo Area

The kingdom of the Barolo lies in the Langhe Hills, south of the town of Alba, at the end of a helter-skelter journey through narrow valleys that descend from the south, past majestic slopes carpeted with vineyards, toward hilltop villages that loom up through the autumn mists. The Barolo region fans out like a hand, with three, mainly north-south facing valleys. Starting in the east, there is the Serralunga d'Alba Valley, then the Barolo Valley, and lastly the valley toward the west that lies at the foot of the village of La Morra. The rivers that form these valleys meet in Gallo d'Alba, gateway to the appellation, before joining up with the Alba le Tanaro that unites with the Pô north of Alessandria.

The Barolo terroirs consist of successive layers of eastward-sloping Miocene terrain. Thus to the west of the village of Barolo, the predominant formations are argillaceous-limestone tuffeau* and Tortonian sands, whereas to the east these gradually sink beneath the white marls and sandy-limestone of the Helvetian. The latter terrain only appears at the summit of hills, in the center of the appellation, and on the vineyard slopes to the east. This is the kingdom of the Nebbiolo, the sole grape authorized for the production of the Barolo. Broadly speaking, Nebbiolo wines from the lighter Tortonian terrain (La Morra sector) are aromatic, supple, and very fruity, while those from the more compact terrain supported by Helvetian marls combine stronger structure and tannins with impressive aging potential.

The quality of the wine however does not depend on the soil alone. The temperamental Nebbiolo is exclusively cultivated in the middle and upper parts of well exposed south- and southwest-facing slopes. The lower sections are reserved for the white grape variety Arneis, while sites with other aspects are planted with the Barbera and the Dolcetto that produce first-rate wines with a more easy-drinking style.

Asti Spumante

As we have just seen, Barolo originates from a sloping basin formed from the confluence of three rivers that meet in Alba and discharge into the Tanaro. Likewise, the hilly terrain of the Asti production area is crossed by a group of rivers that come together around Alessandria. The largest of these rivers, the Tanaro, marks the northern limit of the zone, while the Bormida di Spigno borders it to the southeast. Flowing between them in a northeasterly direction is a group of normally inoffensive but occasionally devastating mountain torrents. These are the rivers Belbo, Tinella, Boglione, and Bormida di Millesimo. After heavy rains, the Belbo in particular has been known to wreak serious damage to cellars throughout the valley.

The Asti Spumante region straddles the two natural environments of the Langhe area in the west, and the highlands

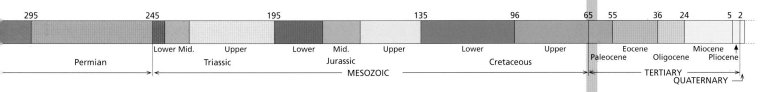

| 295 | | 245 | | 195 | | 135 | | 96 | | 65 | 55 | | 36 | 24 | | 5 | 2 |

Permian		Lower	Mid.	Upper	Lower	Mid.	Upper	Lower	Upper		Paleocene	Eocene	Oligocene		Miocene	Pliocene
			Triassic			Jurassic		Cretaceous								

MESOZOIC   TERTIARY
                                                           QUATERNARY

The village of Barolo, surrounded by softly rounded hills in the heart of the appellation, has given its name to one of the most highly regarded Italian wines.

of the Monferrato region in the east. The landscape in the first sector is distinctly hilly with elongated, almost flat-topped hills, often with ridge roads offering spectacular views of the surrounding scenery. The valleys are deep and narrow. Here, the Miocene terrain produces essentially calcareous soils of whitish marls. The altitude ranges from 523 feet at the bottoms of the valleys to 1,798 feet in Mango and plays a major role in local viticulture. The steep southern slopes often require terracing to aid soil retention but they are richer in limestone than the gentler northern slopes, so favoring good quality Muscat grapes.

The highlands of Monferrato (northeastern part of the region) consist of a series of hills with softer, more rounded contours. The characteristic reddish soils of the region, in vivid contrast with the lush green of the valleys, are derived from more or less sandy Helvetian calcareous sandstone terrain. The altitude here ranges from 523 feet to just 817 feet and the hills are more uniform in terms of both slope and soil composition, which can be quite argillaceous in places.

The Piedmont climate is a mixture of Alpine (continental, dry and cold) and warmer Mediterranean influences. It is characterized by fairly late springs, hot summers and often misty autumns, although average annual rainfall is in the range of 23.6-27.5 inches. There is a marked difference between the Monferrato Highlands, near the warm Alessandria Plain, and the cooler Langhe region where conditions are better suited to the slow-ripening Muscat grape. The grape harvest in Langhe can take place up to three weeks later.

The best Asti terroirs are located in four sub-regions: San Stefano Belbo and Canelli; Calamandrana-Fontanile on the right-bank of the Belbo; and Cassine that overhangs the Bormida di Spigno on the eastern border. The vineyard slopes of San Stefano Belbo and Canelli, overlooking the Belbo River, produce elegant, delicately aromatic wines with a powerful but well-balanced body. Those from the other sectors are stronger and more intensely aromatic, mouth-filling certainly but slightly on the heavy side.

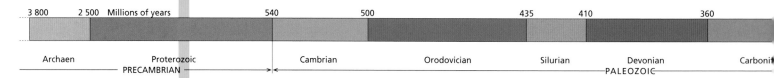

3 800	2 500	Millions of years	540	500	435	410	360

Archaen	Proterozoic	Cambrian	Orodovician	Silurian	Devonian	Carboni
PRECAMBRIAN				PALEOZOIC		

The characteristic hills of the Asti region, in the southern part of the Pô Basin, with steep terrain and Miocene soils on the Langhe side, and softer hills of calcareous sandstone in the highlands of Monferrato.

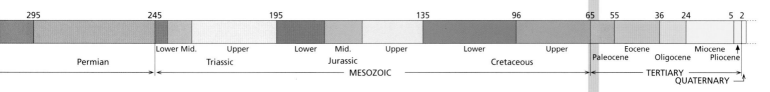

295		245			195			135		96		65	55		36	24		5	2

Permian		Lower Mid.	Upper	Lower	Mid.	Upper	Lower	Upper		Eocene		Miocene	
		Triassic			Jurassic		Cretaceous		Paleocene		Oligocene		Pliocene

MESOZOIC — TERTIARY
QUATERNARY

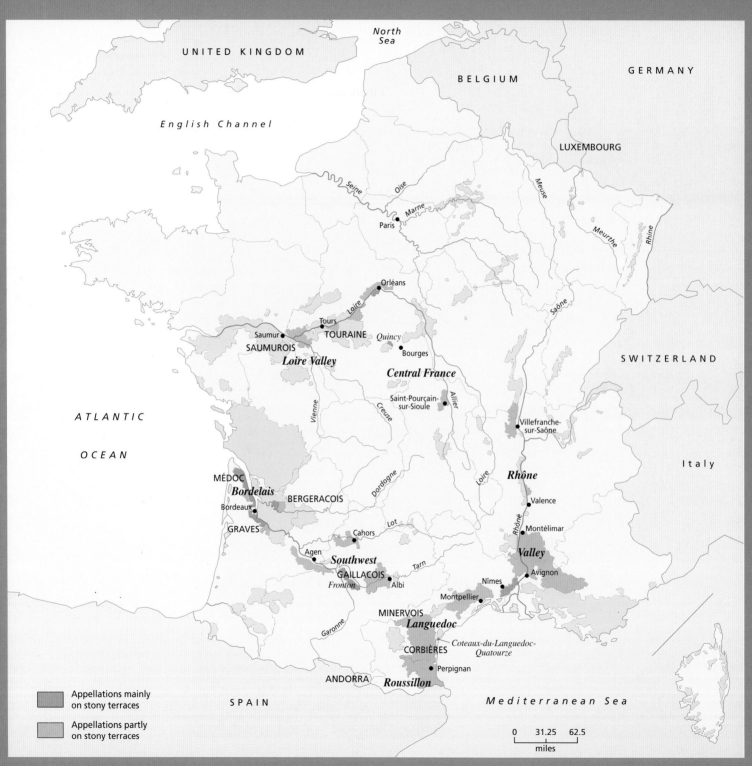

UNITED KINGDOM

North Sea

BELGIUM

GERMANY

English Channel

LUXEMBOURG

Seine

Oise

Marne

Meuse

Paris

Meurthe

Rhine

Orléans

Loire

SWITZERLAND

Tours

Saumur

TOURAINE

Quincy

SAUMUROIS

Bourges

Loire Valley

Central France

ATLANTIC

OCEAN

Vienne

Saint-Pourçain-sur-Sioule

Allier

Villefranche-sur-Saône

Creuse

MÉDOC

Dordogne

Bordelais

Bordeaux

BERGERACOIS

GRAVES

Lot

Rhône

Rhône

Valence

Montélimar

Cahors

Agen

Southwest

GAILLACOIS

Tarn

Valley

Avignon

Fronton

Albi

Nîmes

Montpellier

Garonne

MINERVOIS

Languedoc

Coteaux-du-Languedoc-Quatourze

CORBIÈRES

Perpignan

ANDORRA

Roussillon

Mediterranean Sea

SPAIN

Italy

Appellations mainly on stony terraces

Appellations partly on stony terraces

0   31.25   62.5

miles

Distribution of French vineyards on Quaternary terraces.

Quaternary Terraces

Overhead view of a Quaternary terrace.

Geographers would say that vineyards developed along rivers simply because waterways provided an excellent means of transport. Actually however, terrace soils show certain remarkable characteristics that appear ready-made for the growing of quality vines. Firstly they are rich in stones and coarse, largish gravel (up to a few inches in diameter) that help the soils to warm up quickly by storing heat during the day and returning it to the vines at night. This has a bearing on the rate of ripening which, according to recent research, is a quality factor. The soil warms up deep below the surface, not just at the top.

Stony soils also encourage good drainage by preventing excess water from accumulating in periods of heavy rain. The characteristic gravel mounds of the Médoc and the Graves, a region known for its wet climate, are particularly efficient in this respect. Another important point is that terrace soils, especially the oldest and thus most developed, have undergone a slow process of transformation. The clays that formed in the upper layers have often migrated deeper into the soil, gradually forming a more or less impermeable layer. Rainwater percolates down through the highly porous topsoil and then collects in small pockets on this argillaceous layer. When fine weather returns, a regular moisture supply is drawn up by capillary action to nourish the roots of the vine.

A detailed study of French vineyards on Quaternary terraces shows that distribution varies significantly depending on the river. The biggest concentration is around the Garonne and the Rhône, with more scattered plantings along the Loire and none at all along the Seine and the Saône. The reason lies simply in the relative heights of the massifs from which the rivers originate. Most of the appellation areas with vineyards on Quaternary terraces also cultivate terroirs on other geological formations, especially those overlain with gravel spread. Some appellations are exclusively located on terraces, notably Quatourze near Narbonne, Fronton, Lavilledieu, and Quincy.

The Médoc

The Médoc owes its reputation to an ideal geographical position along the Gironde River that provided an easy means of transport for its wines. Mainly though, the Médoc boasts some exceptional terroirs that seem naturally tailor-made to minimize the effects of the region's relatively high rainfall – a feature that could have prevented quality wine production from this strip of land between the Atlantic Ocean and the Gironde. It only remained for winegrowers to perfect nature's handiwork by selecting grape varieties that were well adapted to the region.

Thanks to its exceptional location, the Médoc represents a world apart within the Gironde viticultural area, with which it is similar in some respects but distinct in others. Similar, because the Médoc like the Gironde is exclusively formed from Tertiary and Quaternary formations. Distinct, because, on the one hand, the flat Tertiary layers are here interrupted by folds and, on the other hand, because they are overlain on the margins of the left bank by thick alluvial gravel.

A World Apart

The geological history of the Médoc in the Tertiary Period is basically not unlike that of the St.-Emilion region. The Médoc remained underwater for longer, notably in the Upper Eocene that saw alternating deposition of Couquèques limestone oyster-shell marls, St.Estèphe limestone, and then fresh deposits of marls. The sea retreated at the start of the Oligocene, and once again Fronsadais molasse underlies the traditional *Calcaires à Asteries* deposited in the Entre-Deux-Mers and Libourne regions by the next marine incursion. These layers were folded at the end of the Tertiary Period by upheavals in the Hercynian basement due to the Pyrenean uplift. This led to the development of two main anticlinal bulges at right angles to the Gironde, one with Listrac-Blaye at its axis, the other passing through Couquèques.

The forces of erosion soon got to work, attacking the *Calcaires à Astéries* that formed the upper part of the bulge and exposing the Mid- and Upper Eocene formations at its core. This softer core material was more severely eroded than the surrounding strata and now provides us with a remarkable relic of inverted relief in the Peyrelebade depression, east of Listrac. Widespread erosion stripped away the traditional *Calcaire à Astéries** along the entire length of the Gironde estuary, revealing the more ancient subjacent formations in the Médoc Grands Crus that now line its shores. So we find Fronsadais molasse in Margaux, and St.-Estèphe limestone (Eocene) in the commune of the same name. The edge of the *Calcaire à Astéries* forms a slight cuesta* that overlooks and runs along the vineyard zone, west of the communal appellations.

The vineyard of Château Marbuzet, southeast of the St.-Estèphe appellation, perched on a steep slope overlooking the Gironde.

3 800	2 500	Millions of years		540	500		435	410		360	
Archaen		Proterozoic		Cambrian	Orodovician		Silurian		Devonian		Carbonife
		PRECAMBRIAN						PALEOZOIC			

By the end of the Tertiary, the landscape of the Médoc as we know it today still had a long way to go. Two major sedimentary processes lasting from the end of the Tertiary to the end of the Quaternary were to give the region its final facelift: gravel spread from the south and east on the left bank of the Gironde, and wind-blown sands from the southwest.

The Gravel Mounds of the Haut-Médoc

At the boundary of the Tertiary and Quaternary periods, when the Pyrenean uplift was at its height, the rivers that flowed down from these mountains brought the alluvial spread that we now find in the upper terrace at the western limit of the communal appellations. These deposits of mainly fine Pyrenean gravel* only cover a large area in Listrac.

At the beginning of the Quaternary, vast quantities of erosional materials were shed by the developing Pyrenees and deposited at their base over the Lannemazan Cone. Forced to deviate eastward toward Toulouse, the Garonne River became the conduit for rivers flowing from the Massif Central. Alluvial deposition now consisted of rock fragments stripped from both massifs. These deposits, known as Garonne gravel, would spread over a series of levels throughout the Quaternary interglacials, or warm periods.

This led to the formation of flat banks of gravel alluvia on different but not always easily distinguishable levels (middle and lower terraces) that ran parallel to the estuary shoreline. The deposits were often thick and composed of bigger pebbles than the spread on the upper terrace. The Médoc terraces, unlike most of the alluvial terraces that we see along rivers today, are not flat but have a very distinctive structure due to the region's location on the edge of an estuary. In the course of the final glacial period (Würm), the level of the Garonne fell by more than 300 feet and small canyons formed at right angles to the river, downcutting the gravel terraces that were more erosion-resistant than the underlying limestones and marls. The edges of these gravelly layers, now in the upper part of the landscape, were slowly worn away by erosion, leaving a succession of rounded gravel mounds.

At the end of the Ice Age, the sea level returned to normal and the small perpendicular canyons between the gravel mounds were banked up by finer elements deposited by the river and the rain. The violent winds from the southwest that marked the end of the Ice Age pushed the sands of Les Landes eastward. Sand piled up in the valleys between the gravel mounds but also spread over the upper terrace, rendering it less suitable for the production of quality wines.

Benchmark Terroirs

So developed the unique Médoc landscape that we see today. The most highly regarded vineyards are established on the mounds of Garonne gravel, at mid-to lower elevations, sometimes on quite steep slopes (in Cos-d'Estournel, for instance). The Tertiary formations supporting these mounds consist of St.-Estèphe limestone in the eponymous appellation, Sannoisian molasse in Pauillac and Fronsadais molasse in Margaux. The streams at right angles to the river (*jalles*) merely indented the tougher limestone formations, leaving large, regular mounds. The softer molasse mounds on the other hand were more severely dissected and now form a collection of gravel islands, especially in the south in Cussac and Margaux. The Recent alluvial fill surrounding the mounds is home to the palus, the uniquely characteristic soils of the Bordeaux appellation on the borders of the Gironde. Here the *palus* defines small valleys at right angles to the ridge and separating the mounds and the communal appellations..There is Estey d'Un, north of St.-Estèphe; Jalle du Breuil between St.-Estèphe and Pauillac; Ruisseau de Juillac between Pauillac and St.-Julien; and Marais de Beychevelle and La Laurine in the Margaux area. These valleys were originally marshy until they were drained by the Dutch in the seventeenth century, leaving arable land not suitable for viticulture. In the rest of the region, the combination of filtering gravel mounds and *jalles* creates a particularly efficient hydric regime that provides the vines with a regular supply of water – a natural advantage that is lovingly preserved by local winegrowers. It is no accident that so many of the Crus Classés, whether on the Gironde side or

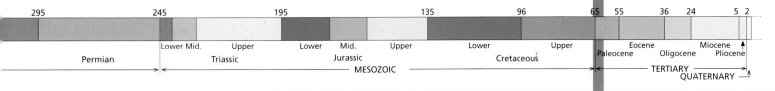

| 295 | | 245 | | 195 | | 135 | | 96 | | 65 | 55 | | 36 | 24 | | 5 | 2 |

| Permian | Lower Mid. Upper Triassic | Lower Mid. Jurassic Upper | Lower Cretaceous Upper | Paleocene Eocene Oligocene Miocene Pliocene |

MESOZOIC · TERTIARY · QUATERNARY

The gravel terraces of the Haut-Médoc, where the vineyards are in competition with the forest (pictured above, Château La Chesnaye).

along the *jalles*, are precisely situated on the edge of these gravel mounds.

The upper, fine gravel terrace, constantly invaded by wind-blown sands and alluvia, was less dissected by erosion and did not develop the characteristic mound shape. The wines produced here are not as highly regarded as their illustrious neighbors. The biggest sector is in Listrac that, like Moulis, lies some distance from the estuary in an area that is home to the largest expanses of limestone and Eocene marls. We find them in a zone shared by the two villages between the upper terrace west of Listrac and the mounds of Garonne gravel east of Moulis. The wines produced here are exceptional, although not quite as excellent as those from the vineyards nearer the Gironde. All of these formations are an ideal growing medium for the Cabernet Sauvignon that continues to account for the bulk of plantings, even though it is officially in decline. It ripens more slowly than the other Gironde cultivars and is particularly fond of the warm, gravel soils.

Listrac and Moulis.

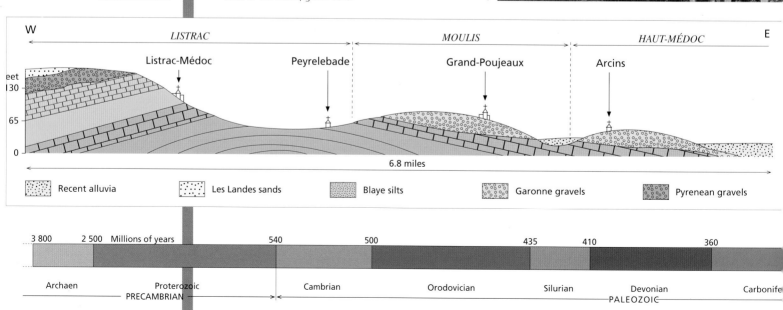

W		LISTRAC		MOULIS		HAUT-MÉDOC	E

Listrac-Médoc — Peyrelebade — Grand-Poujeaux — Arcins

feet
130
65
0

6.8 miles

Recent alluvia	Les Landes sands	Blaye silts	Garonne gravels	Pyrenean gravels

3 800	2 500	Millions of years		540	500		435	410		360	

Archaen — Proterozoic — Cambrian — Orodovician — Silurian — Devonian — Carbonife

PRECAMBRIAN — PALEOZOIC

*Château Latour,
southeast of Pauillac,
is located on the
gravel mounds closest
to the Gironde.*

*From Margaux to the
Blaye region.*

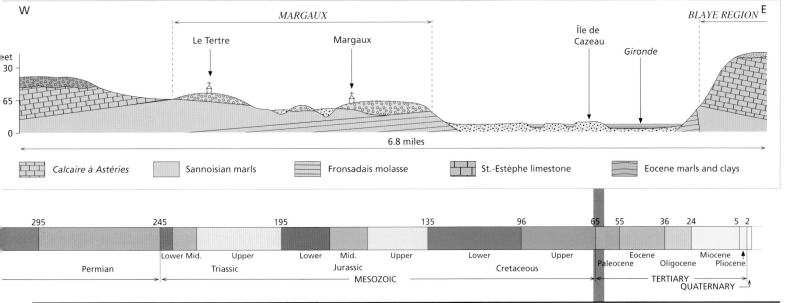

Calcaire à Astéries	Sannoisian marls	Fronsadais molasse	St.-Estèphe limestone	Eocene marls and clays

The Cabernet Sauvignon

Some of the greatest red wines in the world are based on this highly successful cultivar that enjoyed international expansion throughout the twentieth century. The Cabernet Sauvignon is now planted in the United States, mainly in California, as well as in Chile, Argentina, Australia, South Africa and Western and Eastern Europe (Spain, France, Russia, Rumania, the former Yugoslavia, and especially Bulgaria). In France in particular, it now accounts for five times the area it covered 40 years ago.

The Cabernet Sauvignon is at home in a variety of terroirs and climates and owes its expansion to two main factors. Firstly, it is a hardy, very easy-to-grow cultivar. Late budding means it is less prone to spring frosts, and tight clusters of thick-skinned, firm berries make it more resistant to rot, especially in humid climates. These firmer berries may also be picked mechanically without affecting the quality of the wine. Secondly, the Cabernet Sauvignon generally gives high quality results, which can be quite outstanding in some regions. Its performance record, wherever it grows, is far more consistent than that of the Pinot Noir, for instance.

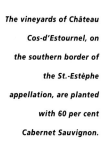

The vineyards of Château Cos-d'Estournel, on the southern border of the St.-Estèphe appellation, are planted with 60 per cent Cabernet Sauvignon.

Grape harvesting on the Los Vascos estate in Peralillo in the Rapel Valley, Chile. The Cabernet Sauvignon is particularly well adapted to the Chilean climate, where cool nights alternating with warm days promote good color development in the grapes.

A Paradox

Due to its easily recognizable organoleptic profile, the Cabernet Sauvignon is currently attracting keen interest as a source of single varietals. Regardless of the terrain or climate, most Cabernet Sauvignon wines have an unmistakable aroma of bell peppers that may be more or less marked. This aroma, known to botanists as aristolochia and to chemists as pyrazine, is common to all grapes in this family (the Merlot, Cabernet Franc, etc.) but is always most pronounced in the Cabernet Sauvignon. In one particular group of terroirs however, this characteristic aroma is noticeably, and paradoxically, absent. This happens to be the very areas where the Cabernet Sauvignon achieves its finest results: on the gravel terraces of the Gironde and especially those of the Médoc. Here, the vegetal notes of bell pepper give way to intensely fruity nuances of ripe black currants, cedar, spices, and sometimes licorice. According to celebrated wine taster Pierre Coste, there is a fundamental distinction between the Cabernet Sauvignon's response to different terroirs. He distinguishes between "fruit" wines and "vegetable" wines, suggesting that those in the latter category are exclusively produced by Cabernet Sauvignon vines in unsuitable terroirs.

When the Grape Yields to the Terroir

The Cabernet Sauvignon's response to its terroir is entirely unlike that of the Pinot Noir. It yields good quality wines with unmistakable character virtually wherever it grows. Another difference between these two major cultivars is that the Cabernet Sauvignon is hardly ever vinified alone but nearly always blended with other grapes. In the Médoc, although it is by far the dominant variety, it is blended with the Merlot, Cabernet Franc, Petit-Verdot, and Carmenère. So the French maxim that " the Cabernet Sauvignon dominates in the Médoc and the Merlot dominates in the Libournais" really applies mainly to the communal appellations. Looking elsewhere in the Médoc peninsula (Médoc and Haut-Médoc AOCs), the Cabernet Sauvignon barely accounts for half the appellation plantings.

It is all to do with the terroir. The Cabernet Sauvignon flourishes in the Quaternary gravel terrace mounds along the Gironde, mainly concentrated in the communal appellations of Margaux, St.-Julien, Pauillac, and St.-Estèphe, and to a lesser extent in Listrac and Moulis. The argillaceous-calcareous terroirs in the rest of the Médoc area tend to favor the Merlot. For this late-ripening variety, the warm, quick drying soils of the gravel mounds of the Médoc are ideal. With their privileged hydric regime, designed by nature and painstakingly improved by generations of winegrowers, they provide a growing medium that is perfectly adapted to the Merlot's requirements.

The soil mechanics of the gravel mounds, and how they provide the vines with a regular water supply, has been clearly demonstrated by Professor Gérard Seguin of the University of

Bordeaux. The best Médoc terroirs promote deep rooting in subsoil that remains moist year-round, so protecting the vines from summer drought. Because they are highly porous and well drained, the soils are also quick to shed excess water, a particular advantage in a region prone to heavy rain. What Professor Seguin did not describe were the characteristic nuances of wines from the different communal appellations. The Margaux wines are fine and delicate, whereas those from St.-Estèphe are more rustic and robust and those from Pauillac combine tremendous power with unique elegance. The St.-Julien wines meanwhile seem to synthesize all of these nuances. The reason for these distinctions remains unclear. They could be due to variations in gravel-mound composition, to the influence of the supporting strata (Fronsadais molasse in Margaux, St.-Estèphe limestone in the eponymous appellation) or to the subtle differences in vineyard techniques from one village to another. Or indeed, to a combination of all three factors – no one really knows.

Cold, Unfavorable Terrain

The Cabernet Sauvignon is planted in gravelly terrain throughout the Gironde. We find it not only in the Graves of course, where it is level-pegging with the Merlot, but also in the gravelly argillaceous Pliocene terroirs of the Entre-Deux-Mers region. St.-Emilion also has plantings of the Cabernet Sauvignon, on the gravelly strip of the upper terrace of Pomerol that extends into the west of the commune, notably in Château Figeac, where it accounts for 35 per cent of plantings. It is much less common on the southern edge of the plateau.

Due to its late budding habit however, the Cabernet Sauvignon has also been planted in colder, more fertile terrain, especially in the Entre-Deux-Mers region. These conditions bring out its grassy, varietal characteristics that can develop into a flaw when yields are too high or the grapes are insufficiently ripe. Vineyards throughout the Gironde increasingly prefer the Merlot to the Cabernet Sauvignon, which is in decline for a number of reasons, partly because of its susceptibility to Eutypa dieback but also because it produces lower-alcohol, slower-maturing wines. Médoc winegrowers are sure to be considering their options in this respect.

Outside the Gironde

In France, the Cabernet Sauvignon is cultivated in the southwest, but also in the Loire Valley and in Provence – everywhere in fact, except for the northeast. Among the Loire Valley appellations, the Touraine cultivates the Cabernet Sauvignon and so especially does Anjou (for the production of Rosé d'Anjou). The Saumurois has much lesser plantings. Practically all of the Loire Valley red wines have that characteristic vegetal nuance of the Cabernet Sauvignon. The only exception are those from Brissac in the Aubance area that bear all the marks of a great terroir. Here, shallow, stony soils resting on Paleozoic schists bring out the Cabernet Sauvignon's finer side. It loses its grassy undertones to take on a peppery bouquet of licorice, lush, ripe fruits, and sometimes the forest floor. The Cabernet Sauvignon plays a much more minor role in Provence, although it has become a feature of some vineyards such as Château Vignelaure. Curiously, when producers in the Baux-de-Provence region were granted appellation status, they asked to be excluded from the *encépagement* (a wine-making term referring to the relative proportions of grape varieties making up a particular blend) so as to focus on wines based on traditional, regional varieties. Could this be the beginning of a new trend?

In the New World

Looking elsewhere in the world, the Cabernet Sauvignon turns in a particularly distinguished performance in California, especially in the temperate vineyards of the Napa Valley. In 1976, a blind tasting by a panel of famous French wine-tasters caused a sensation by ranking a Cabernet Sauvignon from the Stags Leap Wine Cellars (supported by volcanic soils and fluvial sediments) alongside the greatest Bordeaux crus.

Spain also has plantings of the Cabernet Sauvignon, most notably in the outstanding Torres vineyards in Catalonia. In Chile, the Cabernet Sauvignon yields some remarkable wines in the Central Valley region where stony soils derived from weathering of the Andean cordillera and alternating warm days and cool nights provide the grapes with ideal conditions. The Terra Rossa limestone soils of the Coonawarra vineyards in Southern Australia also produce some excellent Cabernet Sauvignon wines (*cf.* p. 188).

Château Gruaud-Larose, produced from vineyards overlooking the Beychevelle marshlands south of the St.-Julien appellation area, is predominantly blended from the Cabernet Sauvignon (around 60 per cent) and lesser quantities of the Cabernet Franc, Merlot, Malbec, and Petit-Verdot.

The Graves and Sauternes Areas

The Médoc, in the Aquitaine Basin, boasts alluvial deposits from the Pyrenees and the Massif Central alike. Not to be outdone, the borders of the Garonne River and its tributaries also feature an impressive cortege of stony alluvia. It is here, on the left bank of the river, that we find the celebrated Graves, a wine area that for many needs no introduction. Extending from the southeast of Langon to the suburbs of Bordeaux, this is one of the few French appellations to be named after its terroir.

The vineyards of Château Rayne-Vigneau, on the Bommes plateau above the Ciron Valley, are planted in croupes of sand and gravel.

The first thing that strikes you about the Garonne is the asymmetry of the right and the left banks. The steep cliff on the right bank is in sharp contrast to the rolling, tiered slopes of the left bank, where four or five levels of sometimes ill-defined terraces stretch over a band of ancient alluvia up to eight miles wide. Toward the south, it represents a continuation of the Médoc terrace deposits.

A Three Tiered Structure

As in the Médoc, the vineyards here have the advantage of alluvial soils although regional geology is actually supported by three main formations:

• Tertiary bedrock, identical to that in the Entre-Deux-Mers area, with *Calcaire à Astéries** at the bottom, then gray Oligocene clay overlain by Lower Miocene sandstone. The Garonne dug its valley out of these formations, right down to the *Calcaire à Astéries* that is visible in places, especially in Barsac but also on the flanks of the tributary valleys.

• Quaternary alluvia found mainly on the left-bank, on the slopes of the Garonne River Valley. The process of deposition was relatively complex and varied from one end of the Graves to the other.

• The wind-blown sands of Les Landes. As in the Médoc, these migrating sands had a marked influence on the region's western boundary.

Gravel Mounds

At the boundary of the Tertiary and Quaternary periods, the Garonne flowed much farther west, through the middle of what is now the forest of Les Landes, toward Guillos, Saucats, and Cestas, some nine or 12 miles from its present-day course. The alluvia it deposited then on the upper terrace (now largely overlain by the sands of Les Landes) were identical to those in the Médoc, Listrac and to the west of Arsac and Pauillac. Tectonic disturbances within the basement created waves in the sedimentary covering. After flowing northward, the Garonne bent east-west at the level of Bordeaux. This meander section of the river then headed back southward as far as the Gat Mort stream, south of the Pessac-Léognan area, where it deposited strips of alluvia that also faced east-west. In the Quaternary, with each interglacial or "warmer" period, the Garonne, which now flowed in a rectilinear fashion northward, drew gradually closer to its present-day course. In the process, the alluvia were deposited at lower and lower levels, closer to the river as it is now. South of Bordeaux, the Garonne cut the upper terrace into strips, reshuffling its constituent elements.

3 800	2 500	Millions of years		540	500		435	410		360

Archaen	Proterozoic	Cambrian	Orodovician	Silurian	Devonian	Carbonife
PRECAMBRIAN			PALEOZOIC			

Château Haut-Brion,
one of the few estates
to have survived
encroaching urban
development in the
Pessac-Léognan AOC.

This produced gravel mounds that, as in the Médoc, became more pronounced in the course of the last glacial period due to the valleys that formed between them.

Pessac-Léognan: The Best of the Graves

Geological evolution created two quite different landscapes on the left bank of the Garonne. From the suburbs of Bordeaux to the Gat Mort stream stretches a green hilly area shared by woods and vineyards. The often-steep gravel mounds, with their natural drainage perfected by generations of winegrowers, make ideal vineyard sites.

This is the heart of the Pessac-Léognan communal AOC where we find the best-classed growths of the Graves area. First among them is Château Haut-Brion, entirely surrounded by the sprawling urban development that increasingly threatens the Graves. The vineyard's other major enemy is the expansion of gravel quarrying activities in the Gironde area. The terroir produces red wines from the Cabernet Sauvignon, Cabernet Franc and Merlot, plus white wines from a blend of the Sauvignon Blanc and Sémillon. The red grape varieties prefer the stony upper part of the slopes, while the white varieties are mainly found in the cooler, more argillaceous sections lower down. Graves white wines are unquestionably the best in the Gironde with their full-bodied, delicate bouquet of toast, citrus, exotic fruits, and hazelnuts and masterful combination of richness and verve.

The Sauternes Area: Gravelly-Sandy Terraces

In sharp contrast are the low hills of the southern part of the Graves where there are no river-cut terraces. Here the terrain slopes gently down from the forest of Les Landes to the river. It is not always easy to see where one terrace finishes and another begins. Vines are no longer grown in monoculture and share the land with the forest. The great dessert wine appellations, notably Barsac and Sauternes, are located in the southern part of the area. Driving from Bordeaux to Langon on the Route Nationale 113, you get your first glimpse of the Sauternes appellation just before Langon where the slopes start to face northward. Sauternes lies in the southern extension of the Graves and slopes down to the Garonne River on the same geological formation. It extends across five communes: Sauternes, Bommes, Fargues, Preignac, and Barsac. Of these, only Barsac also has the right to label wines from its terroir under the Barsac AOC label.

Both of these sweet wine regions lie on *Calcaires à Astéries* bedrock. Largely overlain by Quaternary alluvia, the *Calcaire* only outcrops along the banks of the Ciron River that has dug its channel through this limestone and flows from the Landes to join up with the Garonne between Preignac and Barsac. The *Calcaire à Astéries* is particularly in evidence in Barsac where the overlaying formations have been completely stripped away by river erosion.

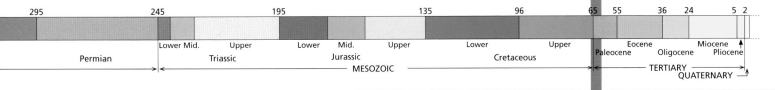

| 295 | 245 | 195 | 135 | 96 | 65 | 55 | 36 | 24 | 5 2 |

Permian | Lower Mid. Upper Triassic | Lower Mid. Upper Jurassic | Lower Upper Cretaceous | Paleocene Eocene Oligocene | Miocene Pliocene

MESOZOIC | TERTIARY QUATERNARY

Château d'Yquem, birthplace of legendary wines born of botrytized grapes.

The largest part of the Sauternes vineyard rests on ancient gravelly-sandy terraces formed by the Garonne and Ciron rivers. These beds of ancient alluvia have been downcut by the rivers to form *croupes* and valleys on three levels:

• The gravelly-argillaceous upper terrace, at an elevation of 228-260 feet on the southwest border of Les Landes.

• The middle, less argillaceous terrace (of the Günz and Mindel glacial periods) formed from coarser gravel. The renowned Château d'Yquem vineyard extends over these first two levels, with the majority of the Grands Crus being located at an elevation of 130-228 feet.

• The lower terrace (of the Riss glacial period), composed of isolated islands mainly in Preignac and in the lower part of Barsac.

The Barsac commune looks quite different on the opposite bank of the Ciron where the *Calcaire à Astéries* forms an area of low, flat, somewhat karstic terrain (49 feet) overlaid by a veneer of coarse argillaceous sands. Running across it like partitions are small, white dry-stone walls built by generations of wine-growers from blocks of limestone found in the soil. This enclave supports several Grands Crus. The Sémillon and Sauvignon Blanc, sometimes blended with the Muscadelle, express themselves best in the sweet wines of Barsac. In the fall, this basin traps the early morning mists that are thought to arise from the convergence of the cooler waters of the Ciron, freshly emerged from the pine-shaded forest of Les Landes, and the warmer waters of the Garonne. But there is another reason for the particular microclimate in Barsac, namely the sharp bend in the river around Langon and the steep right bank at St.-Croix-du-Mont, both of which prevent the mists from escaping easily. And because the rising sun is hidden behind the hills of St.-Croix-du-Mont, these thin fogs take even longer to clear.

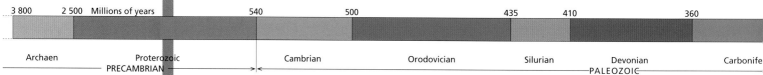

3 800	2 500	Millions of years		540	500		435	410		360
	Archaen		Proterozoic		Cambrian	Orodovician		Silurian	Devonian	Carbonife
		PRECAMBRIAN							PALEOZOIC	

295		245		195		135		96		65	55		36		24		5	2

		Lower	Mid.	Upper	Lower	Mid.	Upper	Lower		Upper		Eocene			Miocene	
Permian			Triassic			Jurassic			Cretaceous		Paleocene		Oligocene		Pliocene	

MESOZOIC · TERTIARY

QUATERNARY

The Sémillon

Plantings of the Sémillon are relatively limited these days and in France, mainly confined to the southwest, especially in and around the department of the Gironde, plus a few dozen hectares in Provence.

The vast majority of white wines from the Gironde are based on a consistent blend of the Sémillon, Sauvignon Blanc, Muscadelle and sometimes the Ugni Blanc and the Colombard. The only part of southwest France not to cultivate the Sémillon is the Béarn, which prefers the Gros and Petits Mansengs varieties and the Courbu. The Pacherenc-du-Vic-Bilh appellation, with its stony soils formed of erosional materials from the Pyrenees, marks a transitional zone planted with both Bordeaux and Béarn cultivars.

The Grape that Loves the "Noble Rot"

The trio of Sémillon, Sauvignon Blanc, and Muscadelle produces a variety of white wines ranging from the sparkling Crémant-de-Bordeaux to the greatest of dry and sweet dessert wines. It is in this last category that the Sémillon really shows its true colors, excelling in vineyards on both sides of the Garonne: from Sauternes, Barsac, and Cérons on the left bank to Cadillac, St.-Croix-du-Mont and Loupiac on the right bank. Throughout these appellations, the Sémillon is the dominant grape and accounts for 70-90 per cent of the finished blend. The Sémillon is in fact perfectly adapted to the specialized production conditions required for these sweet wines. Everything depends on the healthy growth of the mold *Botrytis cinerea* that promotes the spread of the "noble rot". Contrary to what has been suggested by some authors however, the thin skin of the Sémillon berries has nothing to do with it – particularly since this is a thick-skinned grape. What you actually need is the right set of weather conditions to encourage the mold to grow in the first place and, above all, to remain on the berries until the very end of over-ripening. The warm, anticyclonic weather that tends to affect this section of the Garonne is especially favorable, with its early morning mists and hot, sunny afternoons. The berries of course must be thick-skinned enough to survive in these conditions. The *Botrytis* fungus makes them much more permeable, gradually sapping the water from the pulp, concentrating the sugars and transforming their physico-chemical properties. The effect on thin-skinned varieties would be disastrous. The noble rot would turn into gray rot and the berries would perish. But the thick-skinned Sémillon, tougher even than the Sauvignon Blanc, has precisely what it takes to become nobly rotted and produce great dessert wines.

Dessert Wine Terroirs

The Sémillon surpasses itself in the dessert wines of the Sauternes and Barsac AOCs although to what extent its success is due to the soil is hard to say. In fact, the dessert wines of the Gironde originate from three very distinct terroirs. The vineyards of the Sauternes AOC occupy a series of gravelly *croupes* that are the remnants of the terraces formed by the Garonne River. The soils are rich in gravel but also clays. Barsac owns a strip of lower terrace between the *palus* and the railway, but otherwise its vineyards lie directly on clays that have developed from the decalcification of the Stampian *Calcaire à Astéries*, mixed with red sands. On the other side of the Garonne, the vineyards of the Cadillac AOC in St.-Croix-du-Mont overlook the river on a steep ridge of gray clays overlain by Miocene sandstone and capped by Pliocene gravelly clays.

The Bergeracois, the other great dessert-wine-producing region, is also predominantly planted to the Sémillon. Wines from the two premier appellations, Monbazillac and Saussignac, are blended from 74 per cent Sémillon plus varying quantities of the Sauvignon Blanc and Muscadelle (15 percent and 11 per cent, respectively, in Monbazillac wines and 20 per cent and six per cent, respectively, in Saussignac wines).

Sémillon or Sauvignon Blanc?

Vineyards throughout southwest France also produce dry white wines from the Sémillon. They usually consist of a blend of Sémillon and the Sauvignon Blanc, although the proportions of each vary considerably. Unlike the sweet wines that are predominantly based on the Sémillon, some of the dry wines are made entirely from Sauvignon Blanc. The two grape varieties complement each other perfectly. The Sauvignon Blanc is expressive from an early stage, with a well developed, characteristic bouquet and plenty of liveliness due to generous acidity. The more discreetly aromatic but richer Sémillon on the other hand takes two or three years to develop its distinctively honeyed, waxy tones. The Sauvignon Blanc's early maturing qualities explain why it is more frequently replanted than the tardier Sémillon, a grape that is in any case in decline due to its tendency to over production.

On the other hand, the Sauvignon Blanc's distinctive aromas are too overpowering for the sparkling wines (such as Crémant-de-Bordeaux) that are overwhelmingly based on the Sémillon blended with the Muscadelle. Outside France, the Sémillon's greatest successes have been in Chile where it once accounted for 20,000 hectares of plantings but is now seriously in decline. In the course of the past 50 years, Chilean Sémillon plantings have fallen from 17,000 hectares in 1967, to 7,000 hectares in 1985 and just 2,600 hectares in 1996. In Australia, the Sémillon accounts for 3,000 hectares of the area under vine and yields particularly good wines in the Hunter Valley vineyards to the east. Some South African vineyards have extensive plantings of the Sémillon.

Rows of vines at Château d'Yquem (Sauternes).

The Pomerol Area

The modest village of Pomerol has given its name to one of the smallest but most distinguished appellation areas in terms of the complexity of its terroir and the reputation of its vineyards. It is located at the confluence of the rivers Dordogne and Isle and represents the third largest Quaternary terrace zone in the Bordeaux area.

Gravelly terrace soil in Pomerol.

The rambling River Isle flows down from the Périgord in an endless series of east-west meanders, then abruptly changes course after Coutras to head off southward in search of the Dordogne. On the inside of this loop and up to the confluence of the two rivers, the Isle deposited Quaternary alluvia on several levels across an immense, westward-tilting plain. Two streams (the Lavie and the Barbanne) divide the plateau into three, with Pomerol in the south, Lalande-de-Pomerol in the middle, and Artigues-de-Lussac in the north, in the Bordeaux Régionale appellation.

Fronsadais Molasse

All three sections feature the same deposits, but those in Pomerol (the closest to the Dordogne) developed into the finest terroirs. The Fonsadais molasse that was laid bare by erosion at the end of the Tertiary (*cf.* p. 90. St.-Émilion) here underlies the following formations:

• An uppermost ancient terrace (Gunz glacial period) covered

The Pomerol terraces at the confluence of the Dordogne and the Isle.

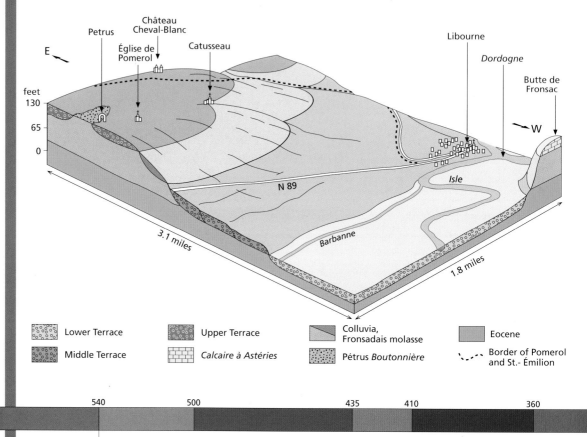

Lower Terrace

Middle Terrace

Upper Terrace

Calcaire à Astéries

Colluvia, Fronsadais molasse

Pétrus *Boutonnière*

Eocene

Border of Pomerol and St.- Émilion

| 3 800 | 2 500 | Millions of years | 540 | 500 | 435 | 410 | 360 |

| Archaen | Proterozoic | Cambrian | Orodovician | Silurian | Devonian | Carbonife |

PRECAMBRIAN

PALEOZOIC

The Pétrus terroir is one of the most astonishing of all, with its swelling clays that provide the vines with a consistent supply of water.

with very stony, more or less thick deposits that overflow into the commune of St.- Émilion (Châteaux Cheval-Blanc, and Figeac).

• Three lower terraces (Riss glacial period) closer to the Isle, with deep soils ranging from fine gravels in the west of the appellation to sands toward the southwest near Libourne. Where the sands meet the subjacent molasse there is a bed of iron slag (known locally as "crasse-de-fer") formed from iron-manganese concretions. These are said to account for the quality of the wines from this particular area, but there is no scientific foundation for this claim.

The wines from the deep, gravelly soils are well built with firm tannins and a bouquet that combines warm notes of empyreuma with licorice, toast, leather, and vanilla. The most complex wines originate from the zone on the border of the St.- Émilion AOC. The wines from sandy soils are willowy and slender, with notes of ripe fruits and vanilla backed by warm hints of cocoa and scorched leather.

The Pétrus *Boutonnière*

Another distinguishing feature of the Pomerol terroir is the famous Pétrus *boutonnière* (buttonhole) in the middle of the upper terrace. This is where the underlying molasse has burst through the gravel cover and now forms a blister at the surface. The soils are rich in so-called swelling clays that provide the vines with a perfectly regulated water supply. The clays consist of microscopic laminae that swell up with water as soon as it rains so sealing off the surface of the soil. An efficient drainage system meanwhile removes any excess water. When fine weather returns, the laminae gradually open up, slowly releasing their water content. If there were not a single vine on the Pétrus blister today, no one would dream of planting one. But thanks to former generations of winegrowers who perceived the ingenuity of nature, these heavy, clayey soils became the source of some outstanding wines.

Pétrus is almost exclusively planted to the Merlot, producing its best-ever results in this little pocket of pure clay. The wines have an earthy, characteristically spicy bouquet with scents of mushrooms, truffles, and red berries (black currant, blackberry). The palate is astonishingly voluminous and rounded with near perfect balance thanks to dense, fleshy tannins that even in the young wines line the mouth with a velvety softness. Some of these wines can age for up to 20 years in the bottle.

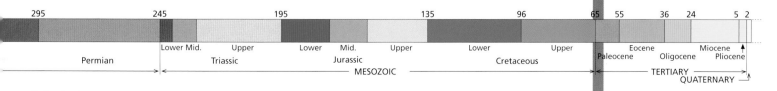

295		245		195		135		96	65	55	36	24	5 2

Lower Mid. Upper Lower Mid. Upper Lower Upper Eocene Miocene

Permian Triassic Jurassic Cretaceous Paleocene Oligocene Pliocene

MESOZOIC TERTIARY

QUATERNARY

The Rhône Terraces from Châteauneuf-du-Pape to La Costière

Gnarled vines emerging from a sea of large, round, polished pebbles have become symbolic of Châteauneuf-du-Pape. But this type of terroir, to a greater or lesser extent, is just as typical of all the communal AOCs of the southern Rhône Valley, including Tavel, Lirac, Vacqueyras, and even part of Gigondas.

On the upper terrace, Villafranchian galets roulés (rolled stones) are sometimes mixed with fragments of the underlying Urgonian limestones.

In this region, the alluvia were deposited in long strips than run more or less parallel to the river. Gone are the gravelly *croupes* of the Gironde. There is no need for significant drainage either because, with low average rainfall (around 25 inches a year) and the ferocious Mistral wind, the vines on these gravelly terraces are more at risk of drought than flooding. The Rhône deposits have served as a worldwide reference in the classification of the different levels of

The hills of Châteauneuf-du-Pape. The Urgonian limestone massif in Lampourdier (on the left of the diagram) protected the Miocene molasse and Quaternary alluvia from fluvial erosion by the Rhône River.

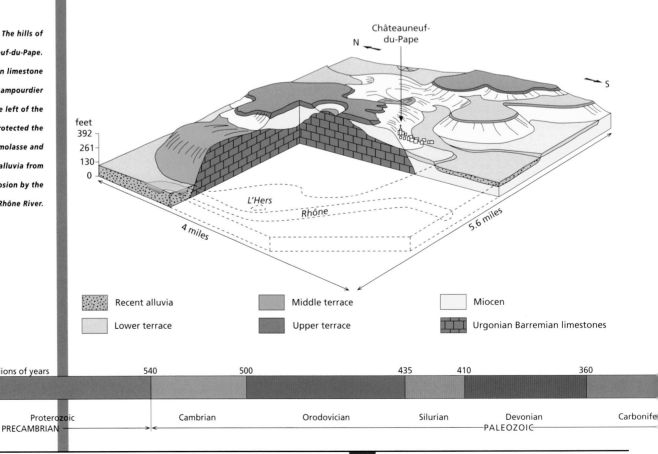

Recent alluvia
Lower terrace
Middle terrace
Upper terrace
Miocen
Urgonian Barremian limestones

Millions of years						
3 800 2 500	540	500	435	410	360	
Archaen	Proterozoic	Cambrian	Orodovician	Silurian	Devonian	Carbonife
PRECAMBRIAN		PALEOZOIC				

Quaternary terraces. They are the product of the alternating cold (glacial) and warmer (interglacial) periods of the Ice Age. All of these levels, from the Villafranchian to the Würm, are found in these southern appellations, in locations that are closely linked to two other classical features of regional geology: very hard, Cretaceous Urgonian limestone (Barremian) and sandy Miocene molasse.

Urgonian Limestones

At the start of the Miocene, before the molasse deposition, this level of the Rhône Valley was blocked off by a series of Urgonian limestone peaks (*cf.* p. 137, Geological setting of the Southern Rhône Valley). These were the origins of the Bois St.-Victor plateau in the west, overlooking Tavel and Lirac,

uplifted and folded by the aftershocks of the Pyrenean upthrust. To the north, the plateau terminates in a sharper fold that on its northern flank plunges almost headlong along a fault that extends from St.-Victor to Roquemaure and Châteauneuf-du-Pape. This is the Mountain of St.-Geniès. The crest of the arch of the fold was worn down by erosion, leaving an impressive rocky barrier that stretches across the Rhône Valley and dominates the landscape around St.-Laurent-des-Arbres in Roquemaure (crossed by the A7 Freeway and the TGV railway line). The series of limestone buttes as you approach Châteauneuf-du-Pape (Lampourdier quarry) were submerged by the Miocene and Pliocene seas that left the Urgonian limestones almost entirely overlain by sandy molasse and sand.

The Rhône Quaternary terraces, showing the lower terrace in the foreground. In the background is the Château de Fines Roches on the edge of the middle terrace.

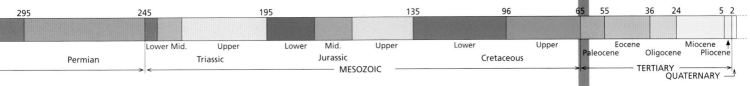

295		245		195			135		96		65	55		36	24		5	2
Permian		Lower Mid.	Upper		Lower	Mid.	Upper		Lower		Upper	Paleocene	Eocene		Oligocene		Miocene	Pliocene
			Triassic			Jurassic				Cretaceous								
					MESOZOIC									TERTIARY				
															QUATERNARY			

The hilltop castle of
Châteauneuf-du-Pape
overlooks the village.
The hills of the
appellation are crowned
by a vast plateau
formed by the upper
Villafranchian terrace.

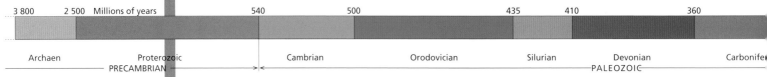

3 800	2 500	Millions of years		540		500			435	410		360	
Archaen		Proterozoic			Cambrian		Orodovician		Silurian		Devonian		Carbonifer
		PRECAMBRIAN									PALEOZOIC		

Stony Terraces on Several Levels

At the height of the Alpine uplift, great quantities of material were washed down by the Rhône that spread sheets of stony debris along the length of its banks in the Villafranchian. The river at that time flowed farther west than it does today and entered the sea at Montpellier. It cut an opening for St.-Laurent-des-Arbres in the rocky barrier of the Mountain of St.-Geniès; and directly behind that, it laid the stony Bois de Clary plateau in the villages of Lirac and Tavel, followed by the Costières de Nîmes plateau. From there, the Rhône veered eastward, around Châteauneuf-du-Pape where it deposited the summital plateau (as found in Cabrières).

In the course of the first Quaternary Glacial Period, as the Rhône bedded into the stony Villafranchian deposits, so the hard Cretaceous limestone buttes came to play a critical role. In the same way that bridge piers obstruct the flow of alluvial scree, so the Urgonian limestone hill of Lampourdier, northeast of the village of Châteauneuf-du-Pape, protected the Miocene molasse and overlying Villafranchian terraces from the onslaughts of the ravaging Rhône. After each Glacial Period, the Rhône cut a progressively deeper and deeper channel, leaving successively lower levels of stony terraces. The limestone butte played a protective role throughout these repeated episodes of downcutting. Today, the bulk of the Châteauneuf-du-Pape vineyards are established on these different levels of stony terraces, from the lowest, Würm terrace that you see as you approach the village from Sorgues to the highest, 392-foot Villafranchian terrace that dominates the hill of Castel-Papale. There are also more minor plantings in the Miocene molasse, the Urgonian limestones that outcrop in the inner zone (Vaudieu, La Gardine), and particularly along the southeastern fringe of the hill of Lampourdier. The soils derived from these formations, unlike those on the Quaternary terraces that conserve moisture in summer, are more drought-sensitive due to rapid water percolation through cracks in the limestone.

The Grenache Noir, blended with the Syrah and the Mourvèdre, expresses itself best of all in the wines of Châteauneuf-du-Pape. The red wines are always richly colored, with a delicately complex bouquet of ripe fruits, truffles, and the forest floor plus wilder, spicier notes. The palate has plenty of force and substance, but retains the same fruity, wild spiciness. The white wines, blended from the Grenache, Clairette, Bourboulenc, and Roussanne, are warm and elegant.

Tavel and Lirac

These vineyards occupy similar formations across the river, facing Châteauneuf-du-Pape. One section of the Tavel vineyards extends northward from the village, over the stony Villafranchian plateau of the Bois de Clary, a formation that is identical to the top of Châteauneuf. The other section stretches westward, into a more recently cleared area of Barremian marly limestones in the valley of Les Vestides. These

Vines growing on the stony Bois de Clary plateau, between Tavel and Lirac.

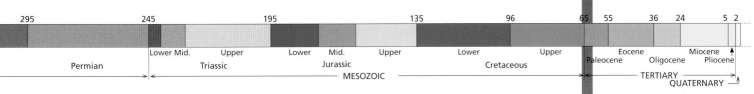

295		245		195		135		96		65	55		36	24		5	2

	Lower	Mid.	Upper	Lower	Mid.	Upper	Lower	Upper		Eocene		Miocene	
Permian		Triassic			Jurassic			Cretaceous	Paleocene		Oligocene		Pliocene
					MESOZOIC						TERTIARY		
												QUATERNARY	

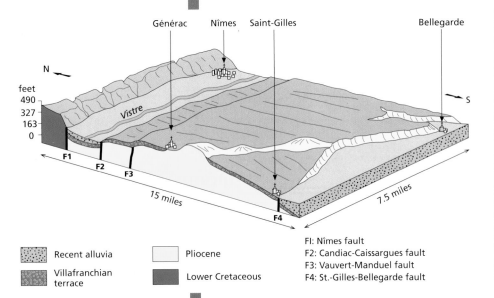

Recent alluvia

Villafranchian terrace

Pliocene

Lower Cretaceous

FI: Nîmes fault
F2: Candiac-Caissargues fault
F3: Vauvert-Manduel fault
F4: St.-Gilles-Bellegarde fault

The Costière de Nîmes.
The Pliocene deposits
were overlain by the
Villafranchian apron
that was then unleveled
by a network of faults,
caused by the collapse
of the Pyrenean-
Provençal range.

are softer than Urgonian limestones but increasingly rich in limestone fragments as you go up the valley toward the Bois-St.-Victor plateau. This is rosé country, producing wines from the Cinsault blended with other traditional regional varieties. The Cinsault expresses itself particularly well in this stony terrain, yielding wines with characteristic finesse and fruit. The formations are more varied in the Lirac appellation and also include Pliocene sands and angular limestone debris along the rocky Urgonian limestone barrier of the Montagne St.-Geniès. Lirac red wines are mouthfilling and rounded but never too strong, with distinctive notes of red berries and stone fruit. With greater maturity, the bouquet develops hints of leather, the forest floor, and licorice. The white wines are intensely floral with a pleasingly fresh, well-balanced quality.

Vacqueyras and Gigondas

The River Ouvèze, notorious for its devastating floods, skirts around the northern end of the Massif des Dentelles de Montmirail, on its way to join the Rhône in Bédarrides. It deposited several levels of terraces, especially in the Mindel glacial stage, including the large strip that remains on the western flank of the Dentelles de Montmirail. These

picturesque slopes are home to the Vacqueyras AOC, southwest of the village itself, and the lower Gigondas vineyards. While they cover a more extensive area of these molassic limestone slopes, the Gigondas vineyards account for a much smaller proportion of Quaternary terraces than the Vacqueyras AOC.

Delicate and generous, Vacqueyras wines have a complex bouquet of red berries and candied fruits, with hints of spice, leather, and game. Those from Gigondas are also very fruity but tending more toward scents of licorice, roasted coffee, and cocoa beans. Mainly though, as you can tell from their powerful structure, Gigondas wines are built to last.

The Costière-de-Nîmes

Formerly known as the Costière du Gard, this appellation area represents the widest expanse of stony fluvial deposits in the whole of France. The vineyards of the Costière de Nîmes cover a vast 12,000-hectares quadrilateral area southeast of the city of Nîmes and north of the Camargue.

Regional geology is fairly straightforward. To the north of Nîmes lies a sparsely planted, garrigue-covered plateau composed of Cretaceous Urgonian limestones that developed along the edge of the Vocontian zone. In the Oligocene, the foreland Nîmes region slumped along the Nîmes fault (F1) that extends as far as Châteauneuf-du-Pape. More acute subsidence in the Miocene, coinciding with the opening up of the Gulf of Lion and the collapse of the Pyrenean-Provençal range, led to step-like ridges separated by faults.

They ran down to the sea like a vast stairway, accumulating Pliocene marine deposits that were subsequently almost entirely overlain by a huge, stony Villafranchian apron. There are outcrops of these Pliocene sandy deposits along a line of hills between Beauvoisin and Bellegarde (the 477-foot Puech de Dardaillon is the highest point of the plateau). The faults between the ridges were constantly reactivated throughout the Quaternary, unleveling the stony Villafranchian apron that forms the Costière-de-Nîmes. The Vistre Valley and the Camargue subsided farther, clearly marking the northwestern

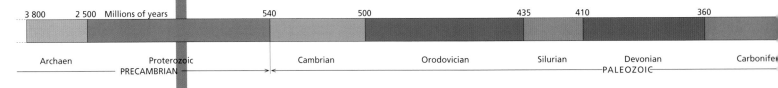

and southeastern limits of the plateau. The remaining edges of the plateau were delimited by the rivers Gard to the northeast and Vidourle to the southwest as they bedded deeper into the sheet of stones and Pliocene sands.

The Costière-de-Nîmes AOC produces red, rosé, and white wines from an area that is entirely defined by the edge of the Villafranchian plateau. For some reason however, the typically "Costière" communes of Comps and Montfrin in the northeast actually form part of the Côtes-du-Rhône AOC. Costière-de-Nîmes wines are blended from the Grenache Noir, Syrah, Cinsault, and Carignan that yield powerful, well-built wines from the southern vineyards facing the sea, and fruitier, more quaffable wines from the northern vineyards, facing the Nîmes garrigues. The Coteaux de La Méjanelle, on the outskirts of Montpellier, used to form part of the Costière-du-Gard AOC but has since become one of the sub-regional Coteaux-du-Languedoc appellations. The Villafranchian terrace also supports the Muscat-de-Lunel AOC, the only Muscat-producing appellation of the Hérault and the Vaucluse with plantings on this type of terrain (rather than limestone).

Tavel produces one of the best rosé wines in France from two principal terroirs: the stony terrace of the Bois de Clary and the marly limestones of the Valley des Vestides.

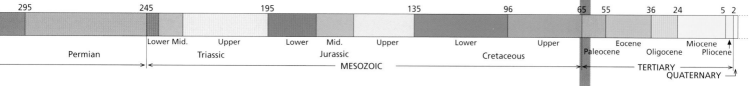

| 295 | | 245 | | 195 | | 135 | | 96 | | 65 | 55 | | 36 | 24 | | 5 | 2 |

| Permian | Lower | Mid. | Upper | Lower | Mid. | Upper | Lower | | Upper | Paleocene | Eocene | Oligocene | | Miocene | Pliocene |
| | Triassic | | | Jurassic | | | Cretaceous | | | | | | | | |

MESOZOIC

TERTIARY
QUATERNARY

The Grenache Noir

Originally from Spain, the Grenache Noir is now cultivated throughout the Mediterranean rim where it expresses itself in a variety of quite different ways. In the Rhône Valley and Provence, it is invariably the leading grape, although traditionally blended with other Mediterranean varieties.

The Grenache Noir thrives in the warm climate and schistose soils of the Maury Valley, in the Roussillon.

In the Languedoc-Roussillon, the Grenache Noir together with the Carignan and the Cinsault form a trio of grapes that produce dry red wines. Plantings of the Grenache Noir have increased despite the Mourvèdre's reprieve and the onward march of the Syrah. In the Roussillon and south of the Aude, the Grenache Noir forms the basis of the *Vins Doux Naturels* and is often vinified alone, especially in the sweet wines of Banyuls and Maury and the best of the Rivesaltes wines. Generally speaking, it performs best in lean, stony soils gorged on sunshine, assuming that yields are kept down.

The Rhône Valley Terroirs

The Rhône Valley is where the Grenache Noir has expanded most and expresses itself best. Geologists distinguish between three broad types of terroir:

• Those on the right bank, especially in the Gard, supported by Upper Cretaceous sandy or marly-sandy limestones that overflow into the Tricastin and the Uchaux Massif on the left bank. Here, the Grenache Noir gives well-balanced, fruity, rich, and rounded wines. Surrounding these hills are Urgonian and Lower Cretaceous plateaus where the Côtes-du-Vivarais vineyards produce fruity, quaffable wines blended from the Grenache Noir and the Syrah.

• Also on the right bank, the terroirs located on the Miocene sands at the bottom of the Valréas and Orange-Vaison-la-Romaine basins. The wines are fruity, light, and supple and include a significant proportion of Côtes-du-Rhône Primeurs wines.

• The terroirs mainly on the left bank, plus a few localized areas on the right bank, that have developed on the Quaternary terraces of the Rhône and its tributaries. The lean soils derived from these formations are a mixture of scree and *galets roulés* (rolled pebbles) in a red argillaceous gangue. This is where the Grenache Noir expresses itself most forcefully, producing generous, robust wines with a spicy bouquet of red berries and prolonged aging potential. The vines are as much in their element in the old established vineyard terraces (Tavel, Lirac, and Châteauneuf-du-Pape) as in the more recent ones (Plan-de-Dieu, for instance).

Mediterranean Terroirs

Elsewhere on the Mediterranean rim, the Grenache Noir is never vinified alone. It may be blended with the Cinsault, which is used to bring finesse and fruitiness to Provençal rosé wines, or with the Carignan and the Syrah, or even, in the Languedoc, with the Mourvèdre. Collioure, that occupies the same steep, Cambrian schistose slopes as Banyuls, blends the Grenache Noir with a touch of Syrah, or Mourvèdre. The wines are warm and vigorous with rich fruit aromas but also with that exceptional finesse and elegance that is characteristic of vines grown in Cambrian schists.

The Grenache Noir, together with the Muscat, are the two principal grape varieties used for the *Vins Doux Naturels*. The Banyuls and Maury VDN are mainly and often exclusively based on the Grenache Noir. Both are equally capable of rivaling the greatest ports, although they originate from quite different terroirs. The coastal vineyards of Banyuls occupy steep slopes of brown Paleozoic (Cambrian) schists, open to sea mists and the Tramontane wind and the ravages of spring and autumn storms. The Maury vineyards nestle inland, tucked between two limestone cliffs that keep their heat in summer. The soils here are derived from black, more or less marly Albian schists (Lower Cretaceous).

These two unique terroirs produce exceptionally aromatic wines from limited yields of Grenache Noir (around 20 hectoliters per hectare). Maury wines have intense aromas of little red berries (cherries, black currant, and blackberries) that with age evolve toward characteristic scents of coffee, tea, cooked fruits, and cocoa. The palate has a fleshy, well-built consistency. Banyuls wines are lighter colored but equally aromatic when young (cherries in eau-de-vie, black currant) while the more mature wines take on notes of coffee, tea, and prunes in eau-de-vie. A bottle-aged Banyuls has a nose of venison and truffles. The Rivesaltes AOC, also in the Roussillon, produces Grenache Noir wines that are different again, from vines planted in stony terraces, Paleozoic terroirs in the Agly Massif and limestones in Tautavel. Some Rivesaltes wines can age for up to 30 years.

In Spain, the Grenache Noir accounts for a significant proportion of plantings in the northeast and center of the country, and ranks as the leading grape variety in several of the northeastern Denominations of Origin (DO) areas. These include Ampurdán, Priorato, Catalayud, Cariñena, and even Tarragona. Blended with the Tempranillo in the Ebre Valley and the Tinto del Pais in Castilla-León, it is never vinified alone but adds characteristically warm, fruity tones to most of the blended wines from these regions.

The small, isolated Priorata DO area, in the Catalonia wine-growing area, produces superb red wines from a blend of the Carignan and Grenache Noir.

The Ancient Basement

Geologists refer to the basement or ancient basement to describe terrain that was violently folded, intruded by granites and metamorphosed in the course of one or several periods of orogeny. The mountain ranges formed in the process were then attacked by erosion and reduced to peneplains that were subsequently overlain by deposits called the sedimentary covering.

In Europe, the ancient basement corresponds to Precambrian and Paleozoic terrain that underwent the Hercynian Orogeny at the end of the Paleozoic Era and was then more or less overlain by Mesozoic sediments. Outcrops of ancient basement rock today occur where the old rocks were not covered by Mesozoic or Tertiary sediments, or where these deposits have been worn away by erosion.

In France, the ancient basement extends across the entire country at varying depths. The principal outcrops are in Brittany, throughout the Massif Central, in the Vosges Mountains, the Ardennes, and at the heart of the new ranges (the Alps and the Pyrenees). It underlies large areas of the Muscadet and Anjou Noir vineyards, and other more scattered plantings along the southern and eastern borders of the Massif Central and the Vosges. In the Iberian Peninsula, the ancient basement mainly outcrops over the larger western sector (southwest Spain and northern Portugal) and supports the Port and Vinhos Verdes vineyards.

In the Southern Hemisphere, the ancient basement is visible throughout a country such as South Africa where the Cape vineyards are entirely established on old rocks and the soils derived from them. A similar situation exists in Australia, although the central region is overlain by Mesozoic and Tertiary deposition. The main viticultural areas have developed on both types of formations.

The vineyards of
the Bellingham estate
in the Franschhoek
Valley in South Africa.

The Muscadet

In terms of geography and viticulture, the Pays Nantais is one of the lower Loire Valley appellations; but geologically speaking, it forms part of the Massif Armoricain. The old rocks of the Hercynian basement remain visible in this Massif that supports the southern part of the Muscadet region. There, the basement rock plunges beneath the Mesozoic sediments that form the Seuil de Poitou, before re-emerging in the Massif Central some 60 miles farther to the southeast.

In the Pays Nantais, the Loire River rambles through a gently undulating landscape worn away by erosion.

This "mountain," unlike most of the rest of France, has a very long history, starting more than two billion years ago in the Precambrian and continuing for more than 230 million years until the end of the Paleozoic. In the course of that time, the region experienced no fewer than four cycles of mountain building, followed by intense erosion.

The Formation of the Massif Armoricain

Of the two cycles that took place before the beginning of the Paleozoic Era (more than 570 million years ago), the most important was the Cadomian Orogeny that saw the formation of a cordillera north of present day Brittany. As we see in the Alps today, the sedimentary covering was folded, compressed (Brioverian schists) and tilted, mainly southward, in nappes that are most famously illustrated by the Champtoceaux promontory in the Ancenis region (Muscadet-des-Coteaux-de-la-Loire AOC).

Of the two cycles that took place in the course of the Paleozoic, the Hercynian Orogeny was the most important. Previously, a marine invasion in the Ordovician had deposited the famous Armoricain sandstones that became very hard and essentially form the present topography of the massif. The Hercynian Orogeny had a significant impact on southern Brittany, especially on the Muscadet region. The Precambrian formations and their Paleozoic sedimentary covering were crushed, compressed, and dislocated along a bundle of faults and folds that starts at the Pointe du Raz and curves in toward the southeast, passing to the southwest of Nantes (Lac de Grand-Lieu region). The level of disturbance was so extreme that geologists have a hard time identifying the origins of these rocks. What's more, each cycle of orogeny brought its share of granite injections, metamorphism, and intense volcanic activity, producing the mosaic of formations that now make up the Muscadet region.

The Impact of Erosion

The end of the Paleozoic Era witnessed the dismantling of the once mighty Hercynian range that was gradually worn away by erosion. Deposition came much later, following a minor marine incursion in the Oligocene, then a second, more major invasion in the Miocene (when the entire Muscadet region was engulfed by the Sea of Faluns) and a third in the Pliocene. These seas brought the *faluns**, then red sands that left thick deposits around the Lac de Grand-Lieu depression. The final stage of deposition came in the Quaternary when cold winds whipped up the silts that cover part of the region.

The Basement Rock of the Nantes Vineyards

The Nantes vineyards are essentially founded on crystalline, metamorphic terrain of volcanic origin dating from the Precambrian to the end of the Paleozoic Era. These formations broadly consist of gneiss, mica-schists, and green rocks

3 800	2 500	Millions of years		540		500		435	410		360

Archaen	Proterozoic	Cambrian	Orodovician	Silurian	Devonian	Carbonife

PRECAMBRIAN — PALEOZOIC

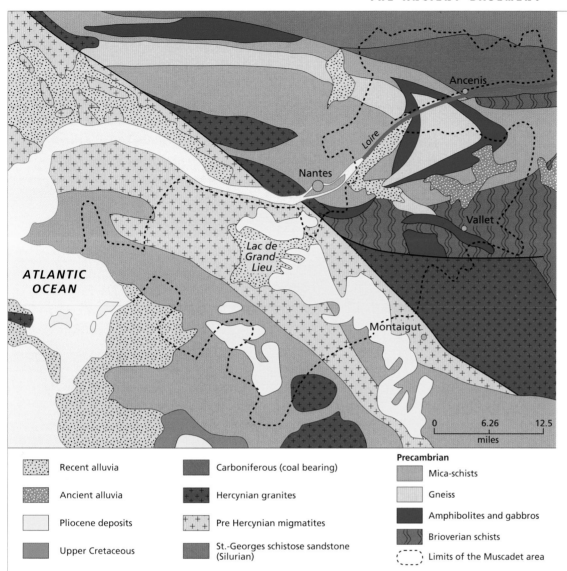

The geological formations of the Muscadet region.

Precambrian

- Recent alluvia
- Ancient alluvia
- Pliocene deposits
- Upper Cretaceous
- Carboniferous (coal bearing)
- Hercynian granites
- Pre Hercynian migmatites
- St.-Georges schistose sandstone (Silurian)
- Mica-schists
- Gneiss
- Amphibolites and gabbros
- Brioverian schists
- Limits of the Muscadet area

0   6.26   12.5
miles

(gabbros* amphibolites* and serpentine*), plus granite to a lesser extent. In the Grand-Lieu region, these rocks are arranged in northwest facing bands, often overlain by essentially sandy but quite heterogeneous Late Tertiary (Pliocene) sediments (red sands, pebbly sands, pebbly argillaceous sands). The soils derived from these rocks are quite varied, ranging from acidic soils formed from coarse-grained granite in Clisson to very basic soils weathered from gabbros in Le Vallet, in the heart of the Muscadet-Sèvre-et-Maine appellation.

Centered on the Muscadet region are two quite distinct but almost entirely overlapping appellations. One is the Muscadet Regional AOC (covering the three sub-regional appellations of Sèvre-et-Maine, Coteaux-de-la-Loire, and Côtes-de-Grand-

Lieu). The other is the Muscadet VDQS appellation (that includes the Gros Plant du Pays Nantais). The Muscadet grape is the Melon de Bourgogne while the Gros Plant is based on the Folle Blanche variety, but the two varieties overlap widely. The Folle Blanche is a later-ripening, more vigorous grape with the potential to be three times as productive, and it is therefore generally planted in less fertile terrain that is quick to warm up (stony slopes). It flourishes for instance in the Pliocene pebbly sandy terrain of the Grand-Lieu region where it produces crisp wines with floral aromas and hints of citrus. The Muscadet by contrast prefers richer, brown soils on gentler slopes with good drainage. It yields dry white wines with a pleasingly fresh palate and mineral undertones.

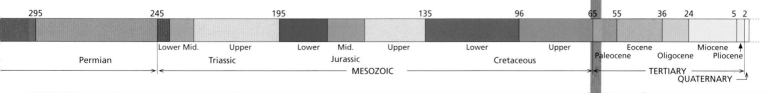

Anjou Noir and the Coteaux du Layon

The dark reddish-brown schistose walls and slate roofs of Anjou Noir ("Black Anjou") are in stark contrast to the light-colored tufa of Anjou Blanc ("White Anjou"). Marking a symbolic limit between the two is the Castle of Angers with its soaring angle towers built of schists and contrasting white tufa. The schists of Anjou Noir have also and most importantly favored the development of the second greatest sweet-wine producing region in France.

The Anjou Noir and Sauternes regions are alike in so many ways that it is difficult to know where to start. Both are located in asymmetrical valleys with a steep southwest-facing scarp on the right-bank overlooking the river, and a gentle slope on the left-bank. Both lie within 60 miles of the Atlantic Ocean. In geological terms however, they could not be more different.

On a Line with the Massif Armoricain

Anjou Noir has exactly the same geological history as the Muscadet region, barring one or two minor but important details. The same Brioverian schists in the Massif des Mauges protect the region on the left bank of the Layon from oceanic influences. In the course of the Hercynian Orogeny, the Ordovician sediments (in the Basin of St.-Georges) were folded, crushed, and transformed into schists that were pierced by volcanic rocks (rhyolites* and spilites*). Meanwhile, the collapsed blocks were invaded by lush vegetation that flourished in the warm, humid climate and would later provide the region with coal-bearing seams.

In the Mesozoic Era, only the Cenomanian Sea made it this far. The topography we see today did not take shape until the aftershocks of the Alpine folding when the right bank was uplifted along the Layon fault that is the continuation of the Montreuil-Bellay fault encountered in the Saumure area.

The soaring towers of the Castle of Angers both symbolize and stand between the Anjou Noir and Anjou Blanc regions.

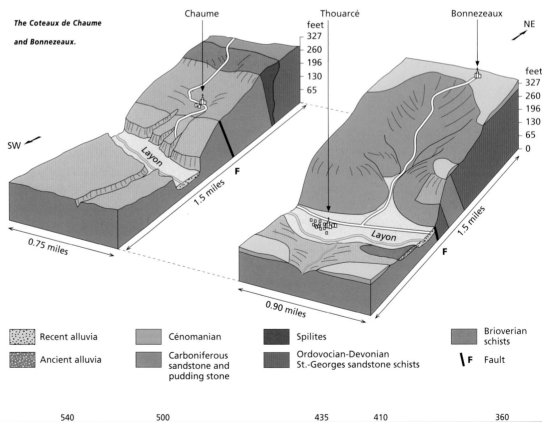

The Coteaux de Chaume and Bonnezeaux.

Recent alluvia · Ancient alluvia · Cénomanian · Carboniferous sandstone and pudding stone · Spilites · Ordovician-Devonian St.-Georges sandstone schists · Brioverian schists · F Fault

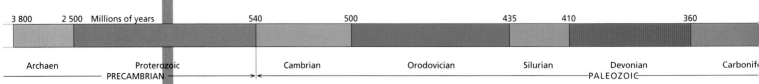

Vineyard Distribution

The terroirs of the Coteaux du Layon run parallel to the main Layon fault in long strips on either side. The sunken block southwest of the fault is formed from green or ochre Brioverian sandstone schists mainly found on the left bank of the Layon but overflowing here and there onto the right bank (Quarts-de-Chaume AOC). In the course of their long period above water, these schists have weathered to produce clayey soils unsuitable for viticulture on flat land. They were also overlain by a veneer of Cenomanian sands and gravels that have survived in places between Thouarcé and St.-Lambert-du-Lattay.

Northeast of the main fault, nearly always on steep slopes on the right bank, we find a narrow band of Carboniferous sandstone and pudding stone*, together with the coal-bearing schists of the Layon Valley that extend from Rablay-sur-Layon to the Loire. In Chaume, more to the northwest where the slope meets the Bonnezeaux fault, the top of the hill is crowned by the schists and sandstone of St.-Georges-sur-Loire, interleaved with spilites* (volcanic rock) and phtanites*. Like the Brioverian schists, these rocks were exposed to intense surface weathering. None of the residues of weathering remain on the slopes, which have come to provide an exceptional growing medium.

The Soil-Noble Rot Connection

These different terroirs yield outstanding sweet wines based on the Chenin Blanc grape. Intense and powerful, they strike a perfect balance between acidity and sweetness. The young wines have rich floral and fruit aromas (white or exotic fruits) that after five or six years take on notes of rare wood, dried or candied fruit, honey, and almonds. The Chenin Blanc grape is highly sensitive to its subsoil. The Coteaux du Layon was included in research carried out by the Institut National de la Recherche Agronomique d'Angers (National Institute of Agricultural Research of Angers) into the relationship between different terroirs and different types of wine. What this study revealed in particular was

The village of Thouarcé at the foot of the hill of Bonnezeaux.

the importance of picking nobly rotted grapes (attacked by *Botrytis cinerea*) in batches so as to bring out the nuances of the terroir. When sweet wines went out of fashion some 30 years ago, growers took to harvesting the entire crop simultaneously, and this practice led to sweet wines that were largely undifferentiated.

Batch picking then became a specified production condition and has since shown that the Chenin Blanc behaves in a particular way depending on the terroir. Also that the *Botrytis cinerea* mold develops differently on vines grown in Carboniferous terrain, Brioverian schists, or St.-Georges schists. Some producers now adapt their methods of vineyard management and harvest planning (number and date of batch pickings) to suit different geological formations – a painstaking approach that calls for the mind of the geologist combined with an artist's hand.

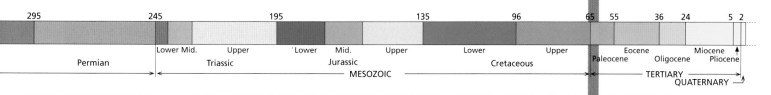

The Chenin Blanc

The Chenin Blanc originated in the lower Loire Valley where it was the making of the fortunes of the Anjou wine-growing region in particular, and still produces its finest results here. It is also widely grown in California and especially in South Africa where it accounts for the most extensive plantings.

The Chenin Blanc thrives in the lower Loire Valley, straddling the Massif Armoricain and the Paris Basin. Being so much at ease here, it becomes uniquely versatile, producing the full gamut, from dry or medium-dry whites to sweet dessert and effervescent wines. All of these wines achieve a consistently high standard of quality and originality, and may in some appellations originate from a single terroir.

Vouvray: Home of Multi-Faceted Chenin Blanc Wines

Vouvray is one of those appellations. The terroir is far from simple and, heading north from the Loire, consists of the following formations:

• Aubuis* soil (Stony, argillaceous-calcareous soil derived from Turonian formations, typical of the Touraine) on the steep slopes of the Turonian.

• Perruches* soil on steep slopes of flinty clays (50 per cent of the appellation).

• Flinty-clay, more finely grained soils on gentler slopes.

There are also considerable variations in local climate due to differences in aspect and distance from the Loire River.

This highly diverse terroir brings out a range of characteristics in the Chenin Blanc. The area known as the Premières Côtes produces exclusively medium-dry and sweet wines. The vineyards

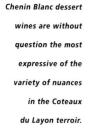

Chenin Blanc dessert wines are without question the most expressive of the variety of nuances in the Coteaux du Layon terroir.

behind yield dry wines while the upper slopes ("Arrières-Côtes") focus on sparkling and effervescent wines. Montlouis, Vouvray's cross-river neighbor, produces the same range of wines from very similar soils enriched by stony alluvial deposits at the confluence of the Loire and Cher rivers.

The remaining Touraine appellations produce dry white wines from the Chenin Blanc, notably the Jasnières and Coteaux-du-Loir AOCs that share the same geology, bordering the River Loir and are known for the quality of their fine, elegant whites. The eastern limit of Chenin Blanc cultivation is Amboise, where the Sauvignon Blanc takes over.

The Sweet Wines of Anjou:
Human Skills in Harmony with Nature

The Coteaux du Layon, in the Anjou area, is home to another optimal Chenin Blanc terroir. The Paleozoic and Precambrian formations of the Massif Armoricain are very much to the liking of the Chenin Blanc, which thrives here in ideal weather conditions, protected by the Massif des Mauges. The slopes are south- and southwest-facing due to the uplift of the ancient basement.

The area is curiously similar to the Gironde (cf. p. 178) and rivals its greatest dessert wines. Here more than anywhere else, winegrowers depend on the batch-picking technique to produce Chenin Blanc wines that express the characteristics of the terroir. Costly and labor-intensive, the technique was abandoned in the mid twentieth century as winegrowers cast doubt on the different nuances in the terroir. Now sweet wines have returned to grace, the batch picking technique has been revived, and those nuances are all the more emphatic.

Chenin Blanc wines from the slopes overlooking the Layon River have an infinite range of expressions due to a combination of the soil and local weather conditions. Bonnezeaux vines are more prone to raisining while those in Quarts-de-Chaume more usually develop the noble rot. However, the development of one or the other phenomenon along the Layon very much depends on annual weather conditions.

Raisined wines certainly take longer to develop, but because like botrytized wines their cellaring potential is virtually limitless, they produce extraordinarily long-lived Chenin Blanc wines.

Across the Loire, the Chenin Blanc puts on another talented performance in the Savennières AOC, yielding dry and sometimes sweet wines of a remarkable delicacy and crispness, from sandstone schists mixed with rhyolites*.

The Chenin Blanc in Hot Climates

The Chenin Blanc has been cultivated in other parts of the world for many years but not with the same success as on the borders of the Loire in the mild climes of Anjou. It is mainly grown in California, usually for medium-dry wines, and especially in South Africa, which has the largest acreage of Chenin Blanc in the world: 28,000 hectares, a quarter of the overall area under vine. Here, Chenin Blanc grapes planted in vineyards with very hot (Olifants River) or Mediterranean climates (Cape district) on Paleozoic sandstone or granitic terrain tend to produce commonplace wines often used for distillation (brandies).

The Savennières vineyards are the continuation of the Coteaux du Layon beyond the Loire River, on the same schists and sandstone schists interleaved with volcanic seams (rhyolites and spilites). Pictured above: Coulée-de-Serrant.

The Port Wine Producing Area

Some appellation wines are not named after the terroir of production but after the place of shipment. Port wines are a good example, being mostly cellared and shipped from Vila Nova de Gaia, on the left bank of the Douro River, facing the city of Oporto, Portugal. The Douro is the soul of Port wine and, with the aid of its tributaries, carved out the valleys and precipitous slopes that are now planted with vines. For centuries, the young wines were shipped by river from their hilly origins down to Oporto and the cellars of Vila Nova de Gaia.

Weather conditions in Oporto, a port town on the Atlantic Ocean, are quite unsuitable for the production of Port wines. On the one hand, the climate is far too humid and, on the other, the granitic, sandy soils are largely unfavorable to viticulture. Oporto in fact lies at the southern end of the Região Demarcada dos Vinhos Verdes, a demarcated region famous for its low alcohol, acidic, tangy wines. The Port demarcated region commences approximately 42 miles upriver, just before Peso de Regua and extends all the way to the Spanish border some 90 miles as the crow flies to the east of the Atlantic coast. Peso de Regua approximately marks the threshold between the rainy, Atlantic weather of the Vinhos Verde region and the drier climate of the Douro valley, home to the port-producing vineyards of the Região Demarcada do Douro (Demarcated Region of the Douro). This is also the beginning of the Cambrian schists that are a characteristic feature of the terroirs in this area.

The Iberian Peninsula

The Port cultivation area needs to be considered within the context of the vast Iberian Peninsula, a region with a geological history quite unlike that of France, even though the chronological landmarks are the same in both cases. The backbone of the region developed in the course of the Precambrian and Paleozoic eras. The Spain-Portugal unit that we see today was formed in the course of the Hercynian Orogeny that took place in the second half of the Paleozoic. When Laurasia (a supercontinent that included North America) collided with Gondwana (another supercontinent that included Africa and South America), the force of the impact between these two great landmasses led to the formation of mighty ranges stretching across the whole of western Europe (such as the Iberian Peninsula, Massif Armoricain, Massif Central, Vosges, and Ardennes).

Orogeny was associated with intense displacement of crustal sedimentary strata, together with igneous intrusions that turned to granite. The surrounding rocks were meanwhile metamorphized by the intense pressure and heat.

The geological history of the Port region from the end of the Paelozoic Era onward recalls that of the Massif Armoricain. The Hercynian ranges were exposed to intense erosion that flattened the entire Iberian Peninsula and laid bare the granitic zones.

By this time, the Spanish Meseta and northeast Portugal (that extends the Meseta toward the east) formed a vast plateau that would remain above water from the end of the Paleozoic Era to the present. This rigid block intruded by granite, known to geologists as a craton, would remain largely undisturbed by the Alpine Orogeny except on its northern and southern borders. The only displacement – but with significant consequences – was a vertical uplift as in the Massif Central. The Douro River then gouged out a deep channel through the schistose and granite formations as it flowed across Portugal from its distant source in Spain (where it is called the Duero) on the borders of Castilla-Léon and La Rioja in the Sierra Cebollera, eventually discharging into the Atlantic around Oporto.

The Douro Terraces

As it crosses the Douro demarcated area (the approved port-wine producing region), the river meanders almost exclusively

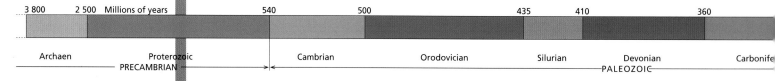

3 800	2 500	Millions of years	540	500	435	410	360
Archaen		Proterozoic	Cambrian	Orodovician	Silurian	Devonian	Carbonife
		PRECAMBRIAN				PALEOZOIC	

The Port vineyards of the Douro Valley and the coastal vineyards of the Banyuls AOC in southern France both produce famous fortified dessert wines from the mutage process (the addition of grape spirit to the musts). Despite their different locations, the Douro Valley and Banyuls are curiously alike from a geological point of view. Both are located on Cambrian schists that have been sculpted by human hand to create a landscape of exceptional beauty.

through the Cambrian schists and graywackes* that form the characteristic Port terroir. Fluvial erosion has had a very particular effect on these schistose formations, producing a highly unusual landscape. Because the layers of schist dip southward, the riverbanks are very asymmetrical in places. The south-facing right bank is less steep than the left bank, especially between Pinhão and Peso de Regua.

These tough, erosion-resistant schists also helped to create the characteristic pattern of pointed hills and steep, narrow valleys, making the landscape along the Douro resemble a grid of tightly interlocked pyramids, some more immense than others.

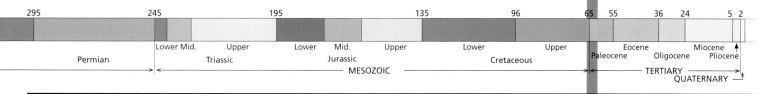

A stairway of precipitous terraces leads down to the Douro River.

Region was approved for inclusion on the UNESCO World Heritage List.

A Microclimate

The Douro terroir is distinguished by a unique set of weather conditions, screened from the rain-laden winds of the Atlantic by the series of peaks that skirts Peso de Regua to the west, with Serra do Marão (4,627 feet) to the north, and Serra de Montemuro (4,519 feet) to the south. Meanwhile, a row of more modest peaks (averaging 3,270 feet) also shield the region from cold north winds. As a result, average annual rainfall is 39-78 inches in the Vinhos Verdes region, 38 inches in Peso Regua, and then gradually decreases eastward, dropping to 27 inches in Pinhão and just 15 inches at the limit of the Douro region on the eastern border with Spain. Average annual temperatures go the other way as the climate gradually changes from Atlantic in the west to Mediterranean in the east. The best Port terroirs lie between the two.

Classification of Vineyard Parcels

In principle, any of the vineyard parcels in the Douro demarcated area can produce port wine. In practice, individual producers are notified of an official maximum yield per hectare, from selected vineyard parcels that are scored on a quality scale. According to this method, the best-quality vineyards at the top of the scale are entitled to produce the largest volumes of wine.

This classification method was developed by Professor Moreira da Fonseca on the basis of a colossal land survey carried out between 1937 and 1945. It takes into account all the major influences on the terroir including: individual parcel location, altitude, gradient, aspect, wind protection, spacing between the rows, varietals planted, age of the vines, the yield, and especially the nature of the soil. Each parcel is scored for one of six categories on a scale of 200 (F)-1,200 (A). Thus in the soil category, schistose parcels (in the majority) score upward of 100 plus points, while transitional terrain scores minus 100 points or lower, granitic terrain minus 250 points or lower and alluvia scores minus 400 points or lower.

The Human Contribution

Wine-growers have worked wonders in these vineyards, sculpting out shapes that emphasize Nature's remarkable handiwork. Witness the pattern of replanting since the phylloxera epidemic. In many places, the traditional small, irregular, horizontal terraces (supporting just one or two rows of vines) have been replaced by wider terraces that follow the contours of the valley and support several rows of vines. Each terrace is separated from the next by schist walls or mounds of earth. More recently, vines have also been planted in rows that run vertically up the steeper hillsides. None of these innovations has disturbed the harmony of the landscape on either side of the Douro. The remarkable landscape we see today is a triumph of human effort and natural processes wrought by the Douro and its tributaries. In December 2001, as a tribute to that combined achievement and following an application by the Portuguese Institute of Port Wine, the Alto Douro Wine

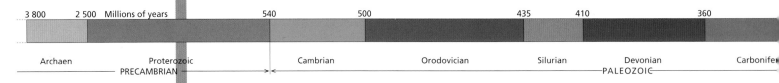

The 145-hectare estate of Quinta do Noval, on the flanks of the Pinhão Valley, near the Vale of Mendiz, produces superb quality ports from 85 hectares of plantings. It is especially famed as the Nacional vineyard, home of the elusive and much coveted Quinta do Noval Nacional, produced from ungrafted vines as was the practice before the phylloxera epidemic.

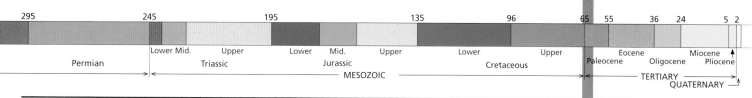

295		245			195			135		96		65	55		36	24		5	2

		Lower	Mid.	Upper		Lower	Mid.	Upper		Lower		Upper		Eocene			Miocene		
Permian			Triassic				Jurassic				Cretaceous		Paleocene		Oligocene			Pliocene	

MESOZOIC   TERTIARY

QUATERNARY

The Vinhos Verdes Viticultural Area

The highly original Minho region, in northwest Portugal, takes its name from the river that marks Portugal's northern border with Spain. Through it flows the River Douro, on its way to discharge into the Atlantic Ocean at Oporto after its journey through the arid vine-terraced gorges of the Upper Douro. Lush green vegetation greets its entry to the Minho, home of the Vinhos Verdes, a far cry from the dry, craggy slopes of the port-wine producing region.

The astonishing differences from one end of the Douro to the other are essentially due to the climate and the soil. The Minho forms part of the Iberian craton – ancient rocks dating back to the pre-Hercynian and Hercynian periods of orogeny. They consist of Paleozoic and Precambrian terrain that remained above water for most of the Mesozoic Era and underwent more pronounced uplifting and faulting in the Tertiary Period.

A Landscape Marked by Granite

Granite is very much the soul of this region and dominates the entire area with the exception of a belt of Silurian-Cambrian schists and slates between Esposendo and São Pedro do Sul, east of Oporto. Otherwise, the soil is derived from weathered granite. The sandy and acidic soil is high in potash but low in phosphoric acid, with a clay content that provides good water retention and protects the plants from drought in summer. Granite is used everywhere you look, from the houses to the posts that support the vine trellises. Conventional wooden stakes would rot too easily in this climate.

A Corrie Open to the Atlantic Ocean

The Vinhos Verdes region resembles a vast corrie or cirque open to the Atlantic Ocean and encircled by peaks. To the north are the peaks that separate the Minho River from the Rio Lima Valley (the Serra de Peneda, 4,630 feet, and the Santa Lucia, 1,805 feet). To the east and southeast lie the peaks of the Serra de Gerès (5,545 feet), de Alvão, do Marão, and de Montemuro. These last three mountains (averaging 4,251 feet) form a neat separation between the Vinhos Verdes region and the Upper Douro region. It is against this rocky barrier that water-laden air masses from the Atlantic Ocean shed the bulk of their water. Throughout the Minho region, average annual rainfall is 59-78 inches. Winter and spring are wet, the summers can be extremely dry, while fall brings frequent rains that can compromise the quality of the grape harvests – the Minho and the Rio Tamega that discharge into the Douro promote dampness. Temperatures are mild and extreme cold is rare.

Six Sub-Regions

Soil and climate contribute to the harmony of the Minho region, with a delightful landscape that seems to radiate every shade of green imaginable. That said, there are subtle differences between each of the six sub-regions that make up the area. Monção in the north, in the shadow of the Inho Valley, is the terroir of choice for the Alvarinho grape (the same grape as the Galician Alvariño in northern Spain) that leaves its signature on some of the best Vinhos Verdes. The red and white wines alike are fruity and light with a slightly sparkling quality.

Farther south, the Lima sub-region in the bucolic valley of the Rio Lima produces blended wines predominately based on the Loureiro. This is also the grape of choice in the Braga area, the most extensive sub-region and home to several independently owned estates now specializing in viticulture. The characteristic vine-growing method of the Minho is to plant vines at the edges of fields with trees providing a natural trellis. In many places, this is now being replaced by more conventional planting methods, but not in the Basto sub-region, in the heart of the upper Rio Tamega Valley. Traditional practices remain the rule here. This is one of the most mountainous but also the most beautiful of areas, producing some of the most characteristic red Vinhos Verdes: light, fruity, often slightly sparkling, and very sharp. Peñafiel is the last of the six sub-regions, with large, modern vineyards centered on the towns of Felgueiras, Peñafield, and Paredes.

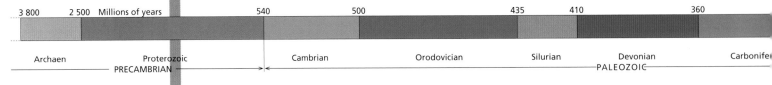

| 3 800 | 2 500 | Millions of years | 540 | 500 | 435 | 410 | 360 |

Archaen · Proterozoic · Cambrian · Orodovician · Silurian · Devonian · Carboniferous

PRECAMBRIAN · PALEOZOIC

Vines in the Vinhos Verdes region traditionally grow at the edges of plots planted with other crops (opposite) or are trained along rows of trees. These days, however, planting in rows is increasingly common.

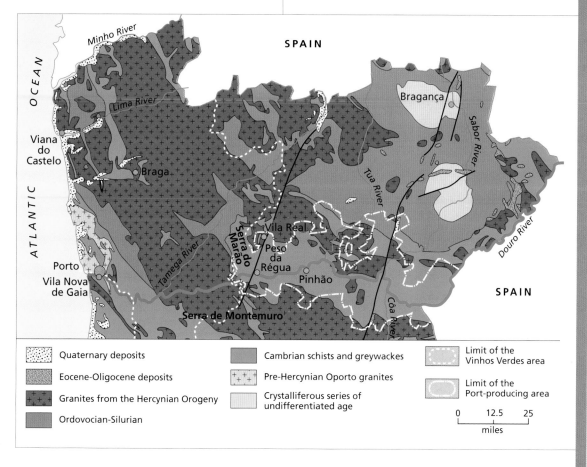

The Geology of the Port and Vinhos Verdes Viticultural Areas.

Legend:

- Quaternary deposits
- Eocene-Oligocene deposits
- Granites from the Hercynian Orogeny
- Ordovocian-Silurian
- Cambrian schists and greywackes
- Pre-Hercynian Oporto granites
- Crystalliferous series of undifferentiated age
- Limit of the Vinhos Verdes area
- Limit of the Port-producing area

0   12.5   25
miles

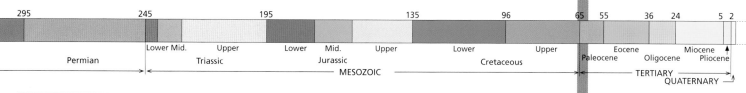

295	245	195	135	96	65	55	36	24	5 2

Permian | Lower Mid. Upper — Triassic | Lower Mid. Upper — Jurassic | Lower Upper — Cretaceous | Paleocene Eocene Oligocene Miocene Pliocene

MESOZOIC — TERTIARY — QUATERNARY

South Australia

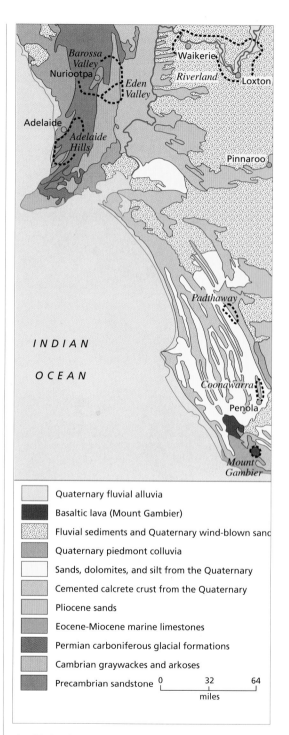

This huge island continent, the size of the United States, has been cultivating vines since the eighteenth century. The vineyards today are mainly established in the southeast of the country, between Brisbane, Melbourne, and Adelaide, with more minor plantings in the southwest, south of Perth. The most significant viticultural area, in terms of both output and reputation, is the Adelaide region in southern Australia.

Geologically speaking, the gigantic island continent of Australia once formed part of a huge block that also included Antarctica and India, which then lay alongside South Africa. It did not become detached from Antarctica until the Tertiary Period, some 60 million years ago, when it started to drift northward (*cf.* p.12) as a single unit. The Australia we see today may be divided into three main geological areas:

• A western sector that is part of a Precambrian craton* formed some of from the oldest rocks on Earth and certainly the oldest in Australia.

• An eastern region consisting of the Great Dividing Range or Eastern Cordillera that was formed in the course of the Hercynian Orogeny (Paleozoic Era). It extends along the Pacific seaboard, from Melbourne to the Cape York Peninsula.

• A central region formed from a vast depression between the two previous sectors, which became filled by Mesozoic and Tertiary marine and continental deposits.

A Wealth of Geological Supports

The viticultural regions of South Australia are supported by a variety of geological formations that include not only ancient rocks but also Tertiary and Quaternary sedimentary rocks and volcanic soils. They are included in this chapter because the oldest and most important vineyards in the State of South Australia are overwhelmingly established on old rocks.

The cream of South Australia's vineyards are located in the

Legend
- Quaternary fluvial alluvia
- Basaltic lava (Mount Gambier)
- Fluvial sediments and Quaternary wind-blown sand
- Quaternary piedmont colluvia
- Sands, dolomites, and silt from the Quaternary
- Cemented calcrete crust from the Quaternary
- Pliocene sands
- Eocene-Miocene marine limestones
- Permian carboniferous glacial formations
- Cambrian graywackes and arkoses
- Precambrian sandstone

0   32   64
miles

The Viticultural Areas of South Australia

Millions of years							
3 800 2 500		540	500		435	410	360
Archaen	Proterozoic	Cambrian	Orodovician	Silurian	Devonian	Carbonife
PRECAMBRIAN		PALEOZOIC					

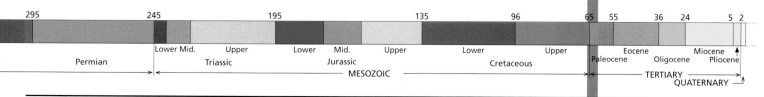

The vineyards of the Barossa Valley extend from Nuriootpa to Lyndooh, in a warm, dry climate in the heart of the Mount Lofty Ranges.

Mount Lofty Ranges, a crescent-shaped range of hills running first north-south then northeast-southwest to the east of Adelaide, extending at their southernmost point into Kangaroo Island. They correspond to the non-submerged part of the Adelaide geosyncline that was formed in the course of the Delamarian Orogeny, the last orogenic period to affect this region some 480-500 million years ago in the Ordovician Period. Today, the deeply weathered range rises no higher than 3,041 feet on the northern side. Flanking it to the west and especially to the south and east are Tertiary and Quaternary fluvial, lacustrine, or wind-blown limestone and sands, plus gravelly deposits on the plains.

The Viticultural Regions
of the Mount Lofty Ranges

The main and also the most famous wine-growing region is located in the Barossa Valley, around the towns of Nuriootpa, Tanunda, Lyndooh, and Angaston. The valley lies on Quaternary sands and gravels surrounded by Precambrian and Paleozoic terrain. The altitude here is modest (882 feet on average) but varies considerably, providing a diversity of vineyard sites with predominantly brown argillaceous sandy soils. The Shiraz (Syrah) is the king of the Barossa Valley and particularly renowned as the basis of Penfolds Grange, the most famous of all Australian wines. This inky-colored wine combines an exuberantly spicy bouquet of violets, licorice, juicy black currant, and rare woods with a dense palate supported by pungent, fleshy tannins.

The neighboring Eden Valley region, southeast of the Barossa, is equally wild and rugged but located at higher altitudes (1,242-1,635 feet) where the cooler temperatures are more suitable for white wine production. The soils are derived from Cambrian mica-schist and range from sandy-silty to argillaceous-silty with a high proportion of gravels, quartz, and other rock fragments. In addition to its celebrated Riesling wines, the Eden Valley also produces first-class wines based on the Syrah and the Chardonnay. There is little change in either terrain or altitude in the Adelaide Hills area, to the south.

295		245		195			135		96		65	55		36	24		5	2

Permian	Lower Mid.	Upper	Lower	Mid.	Upper	Lower	Upper	Paleocene	Eocene	Oligocene	Miocene
	Triassic		Jurassic			Cretaceous					Pliocene
			MESOZOIC						TERTIARY		
									QUATERNARY		

The vineyards of the Henschke estate are located at the heart of the Eden Valley, toward the eastern end of the Mount Lofty Ranges.

The Viticultural Regions of the Plain

The Mount Lofty Ranges are surrounded by vast sedimentary zones. A good example is the Riverland region to the north, along the Murray River that drains all of southeastern Australia. This high-volume producing region traditionally focused on quantity rather than quality, but things have changed. Riverland lies farther inland than the previous areas. The climate is hot and dry and rainfall occurs mainly in winter, making irrigation essential. Water comes either from the river that cuts a deep 160-foot channel through the plateaus, or from water pockets stored in underground layers of clay. The sandy-silty soils in these Pliocene formations frequently overlie deep limestone layers.

The Padthaway region, slightly farther south, lies about 150 miles from Riverland along the dead straight border with New South Wales. Just 40 or so miles from the sea and with no significant mountain ranges to protect it, this area is open to cooler maritime influences. It extends over an immense plain of terrain that is rippled by slight differences in the soils. These are derived from early Quaternary deposits and consist of reddish-brown clays, known as *terra rossa*, over localized limestone strata. The region is mainly focused on white wine production.

Coonawarra

These vineyards some 60 miles farther south have become so successful in the past 50 years that they now occupy the most

3 800	2 500	Millions of years	540	500		435	410	360
Archaen	Proterozoic		Cambrian	Orodovician		Silurian	Devonian	Carbonife
	PRECAMBRIAN						PALEOZOIC	

Harvesting in Coonawarra. The area corresponds to a narrow strip of terra rossa soil barely one mile wide and roughly 15 hectares in area.

expensive viticultural land in all of Australia. The Coonawarra region is located some 40 miles from the Indian Ocean. With no hills to speak of, it enjoys a maritime climate that is cooler and slightly less sunny than Padthaway. Rainfall rarely exceeds 16 inches per annum but the region has abundant water supplies thanks to deep groundwater reserves. The geological formations are much the same here as in Padthaway and along the Limestone Coast, southeast of the Mount Lofty Ranges and Adelaide. This is a region of low-lying terrain (maximum altitude 163 feet) composed of alternating bands of **terra rossa** that run parallel to the coast, and lower zones of deep, black, heavy, and often marshy terrain (Penola, south of the Coonawarra zone, is Aborigine for "Great Marsh"). The area was formed by a complex set of advances and retreats of the sea in the Pleistocene Ice Age.

The Coonawarra viticultural region, about nine miles long and barely one mile wide, is located on one of these bands of **terra rossa**. The soils are derived from sand dunes that were driven inland by the winds. Along the **terra rossa** strip, calcium carbonate shell debris was redeposited downward by run-off water, while the shiny, friable red clays remained on top. Gradually the limestone accumulated in a thick white, cemented layer of variable hardness called calcrete*. The result is a soil with good natural drainage that is very favorable to

viticulture. This type of terrain is slightly higher than the surrounding area of black, heavy, hydromorphous earth that produces exclusively white wines. The red soils of Coonawarra on the other hand are the chosen terrain of the Cabernet Sauvignon and yield the pick of the Australian Cabernets. The wines of Coonawarra strike a subtle balance between aromatic richness and woodiness. Their fruity bouquet, entirely free of those vegetal characteristics typically associated with the Cabernet Sauvignon, bears all the marks of a great terroir.

Our tour of the Limestone Coast would not be complete without mentioning the small viticultural area of Mount Gambier, noted for its volcanic soils. It is located at the tip of the State of South Australia, around the towering volcanoes of Mount Gambier and Mount Shank that covered the surrounding limestones with their lava flows.

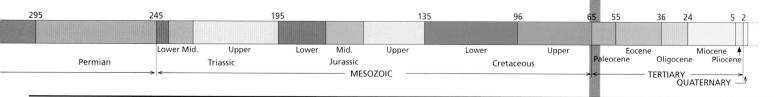

| 295 | | 245 | | 195 | | 135 | | 96 | | 65 | 55 | | 36 | 24 | | 5 2 |

Permian — Lower Mid. Upper Triassic — Lower Mid. Upper Jurassic — Lower Upper Cretaceous — Paleocene Eocene Oligocene Miocene Pliocene

MESOZOIC — TERTIARY — QUATERNARY

South Africa

South Africa is one of the few countries in the world to know precisely when the first vine was planted on its soil and by whom. The year was 1655 and the person responsible was Jan Van Riebeeck, founder of Cape Town. The first harvests took place in 1659. All the same, South African vineyards are still relatively young, while the formations they occupy rank among the most ancient on the planet.

The last major tectonic shift in this part of the world took place in the Upper Cretaceous when Africa became detached from South America and Antarctica. Events after that were confined to folding and erosion of the formations created by the shift, plus igneous intrusions that produced South Africa's fabulous diamond mines.

The Cape Vineyards

Half of the vineyards in South Africa are found in the Cape region, between the Atlantic Coast and the Great Escarpment that rings the high central plateau on which most of the country is built. The growth of viticulture in the Cape is a product of the Mediterranean climate, which in turn is largely due to powerful but opposing currents that flow around the southernmost tip of the African continent. One is the cold, Benguela Current, off the west coast of the Cape. The other is the warm Agulhas Current from the Indian Ocean, off the southwest coast and in False Bay.

The most ancient rocks in the Cape region (the so-called Malmesbury Group) were formed in the Precambrian, between one billion and 600 million years ago. These sedimentary deposits were then compressed in the course of the Namibian Orogeny, and subsequently intruded by granites in the Cambrian. In the process, the surrounding layers were metamorphosed to form schists, phyllites*, graywackes*, and granites. The mountains so formed then came under attack from erosion and were practically razed to sea level. In the second half of the Paleozoic, thick deposits of sandstone and schists (essentially Ordovician in age) accumulated across this eroded surface (Cape sedimentary group).

In the Mesozoic Era and Tertiary Period, the entire continent underwent general uplift and this plateau formed by deposition of sediments came under attack from erosion, starting with the coast. The edge of the plateau gradually retreated to its present location at the foot of the Great Escarpment, overlooking Paarl and Franschhoek to the east. Prolonged erosion eventually exposed the subjacent rocks of the Malmesbury Group, intermingled with sandy residues from weathering of

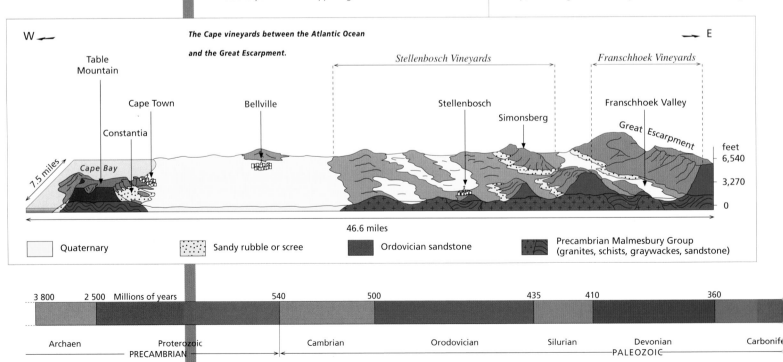

The Cape vineyards between the Atlantic Ocean and the Great Escarpment.

W ← → E

Table Mountain
Cape Town
Constantia
Cape Bay
7.5 miles
Bellville
Stellenbosch Vineyards
Stellenbosch
Simonsberg
Franschhoek Vineyards
Franschhoek Valley
Great Escarpment

feet
6,540
3,270
0

46.6 miles

Quaternary	Sandy rubble or scree	Ordovician sandstone	Precambrian Malmesbury Group (granites, schists, graywackes, sandstone)

| 3 800 | 2 500 | Millions of years | 540 | 500 | 435 | 410 | 360 |

| Archaen | Proterozoic | Cambrian | Orodovician | Silurian | Devonian | Carbonife |
| PRECAMBRIAN | | | | | PALEOZOIC | |

the Ordovician sandstone. A few sectors survived erosion, including Table Mountain, Cape Town's celebrated backdrop that towers over the region from a height of 3,551 feet.

It was on the eastern side of this mountain that in 1679 Simon Van der Stel, one of the Cape's first Dutch governors, established his Constantia estate. Two centuries ago, the Muscat à Petits Grains was planted on these sandy slopes of eroded sandstone from Table Mountain, mixed with small quantities of Malmesbury Group sandstone and granites. Thus were established the sweet wines that became famous throughout the whole of Europe. Nowadays the Muscat has been replaced by the Chardonnay, Sauvignon Blanc, and Cabernet Sauvignon, and the sweet wines have made way for some excellent dry whites.

The Vineyards of Paarl and Franschhoek

There are more vineyards at the other end of the coastal plain, some 30 miles farther east, in the foothills of the Great Escarpment. These sandy slopes of Ordovican sandstone mixed with Namibian granites and gneiss are home to the rolling vineyards of Paarl. Some are located in blind valleys carved out of eroded sandstone, such as the Franschhoek Valley (Afrikaans for "The French Corner") where the vines face the sun in a corrie that opens to the north. This little corner of paradise overlooked by towering sandstone cliffs is where

elephants used to come to give birth, before French Huguenots fleeing religious persecution in the Luberon arrived at the end of the seventeenth century and planted vines. Today they line the valley floor and sides, growing in a mixture of Precambrian granites and conglomerate, mixed here and there with sandy residues from weathering of the surrounding sandstone peaks. The wines from this region are regarded as some of the best in South Africa. They include white wines based on the Sauvignon Blanc and Chardonnay, and reds made from the Cabernet Sauvignon, Merlot, Syrah, and Pinot Noir.

The Stellenbosch Vineyards

Cape wine production is centered on the Stellenbosch region, between Constantia and Franschhoek, where the terrain becomes much more complex. Malmesbury Group schists and granites outcrop in long northwest-southeast facing strips that are intersected by residues of Ordovician sandstone sands, plus Quaternary river alluvia. Recent soil studies have demonstrated that the best terroirs lie around the Ordovician sandstone hills that survived erosion, such as the Simonsberg Mountains. The mosaic of soils in this area suits a variety of cultivars, ranging from the Chenin Blanc (for the production of dry and sweet wines, and spirits) to the Cabernet Sauvignon, Sauvignon Blanc, and Chardonnay.

Above:

Cape Town, set against the majestic backdrop of the Ordovician massif that forms Table Mountain (3,551 feet).

Left:

Stellenbosch was one of the first South African areas to cultivate vines in 1769, preceded by Constantia in 1679.

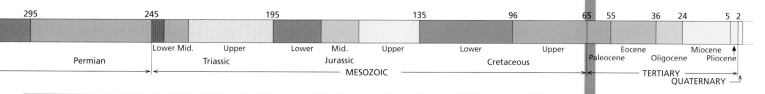

Vineyards in the Foothills of Mountains

Vines thrive on sunny, well-drained slopes commonly found in mountainous zones where the only restrictions to planting are altitude and aspect. The closer you get to the Equator, of course, the higher you can plant your vines.

A geological feature of mountainous zones is that from the moment they were uplifted, they became a target for erosion* that homed in on the highest peaks. The erosional products that were shed in the process piled up in the foothills of massifs before being gradually washed away by rivers.

Material eroded from the highlands was dumped in lowlands and valleys, either through force of gravity or by the movement of water, whether liquid in rivers or solid in glaciers. As this detritus gradually spread across the valley floors, it formed a more or less flat, typically stony growing medium that is highly favorable to vines in temperate zones such as northern Italy, Savoie (France), and Switzerland.

In warmer, drier regions the vine readily adapts to deeper, finer-textured soils. This is the case in the vineyards of the central Andes, both on the Chilean (Central Valley) and the Argentine (Mendoza region) sides, and in California along the coastal peaks of the Coast Ranges. Unlike more temperate zones, these regions have no excess water problems – rather the reverse. Water shortages are the norm, making irrigation essential in many places.

The combination of warm days and cool nights in or around mountainous massifs is another good reason to plant vineyards there, especially in the case of red wines because these diurnal fluctuations help to concentrate the pigments in grape skins.

The Valais vineyards face the sun in the foothills of the Swiss Alps.

Savoie

The Jongieux Cru provides the Altesse variety with a terroir of choice on the western slopes of the mountains of Charvaz and du Chat, where Kimmeridgian marls are often overlain with scree.

The viticultural region of Savoie consists not of a single vineyard but a string of highly distinctive crus, scattered along the foothills of the Alps from Lake Geneva to the Isère Valley. The landscape is very diverse and weather conditions are highly variable. The different terroirs are essentially mountainous but moderated by the presence of Lake Geneva and Lake Bourget.

Les The French Alps are home to the highest peaks in Europe, but they are very open and intersected by deep valleys that clearly mark the divisions between the main mountain blocks and provide easy routes into the heart of the massif.

History of the Western Alps

Throughout the Mesozoic Era, before the Alps started to grow, their present-day location was under siege from a deep sea that overlay the Hercynian basement* with alternating layers of hard and soft deposits. The Oligocene was a period of relative calm, marked by the collapse of the Bresse and Valence regions that became filled with material shed from the surrounding hills. In the Miocene, this collapsed zone was invaded by waters from the Mediterranean that embraced the nascent Alps to the north, so forming the peri-Alpine trough that accumulated deposits of molasse (Rhône Valley, Lake Geneva).

By the end of the Miocene, the uplift of the French Alps was at its climax. The Hercynian basement underwent significant uplift movement with a push toward the northwest. The Mesozoic covering came unstuck and, like a huge sheet, slipped and folded along the northwest side of the basement while the front of the sheet pitched downwards into the Miocene molasse of the peri-Alpine trough.

This Mesozoic covering was chopped up into a series of small, regular blocks that we now know as the Massifs of Bornes, Bauges, La Chartreuse, and Vercors. Between these two Hercynian and sedimentary parts of the Alps, the soft Jurassic layers at the base were scooped out to form a deep trough (Mégève, Albertville, Pontcharra, Grenoble).

Glaciers in the Quaternary Glacials reached all the way to Lyon. They widened the nascent valleys and deposited moraine* over the Miocene molasse in the plain (plains of Thonon-les-Bains, Geneva, and west of Annecy). Erosion was particularly fierce in these mountain massifs and ranged from fluvial erosion with stony terrace deposition* to talus formation and

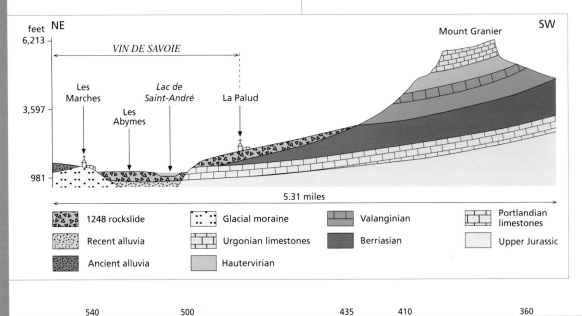

The terroir of the Cru Abymes.

1248 rockslide	Glacial moraine	Valanginian	Portlandian limestones
Recent alluvia	Urgonian limestones	Berriasian	Upper Jurassic
Ancient alluvia	Hautervirian		

Millions of years

3 800	2 500		540	500		435	410		360

Archaen	Proterozoic	Cambrian	Orodovician	Silurian	Devonian	Carbonife
PRECAMBRIAN		PALEOZOIC				

the development of alluvial fans*, spreading out from the points where torrents entered valleys.

Mainly Quaternary Soils

The geological history of this part of the Alps helps to explain the present-day configuration of the different massifs. Their physical components however – Paleozoic terrain on the Hercynian basement, Mesozoic terrain on the sedimentary covering, plus Miocene molasse – provide hardly any growing mediums for the vines. What they prefer by far in this mountainous environment are stony, detrital soils, particularly glacial moraines, scree fans, and fluvial terraces. A handful of vineyards have also become established on the Mesozoic terrain of the sedimentary covering, such as the limestones and marly limestones of the Jurassic (Kimmeridgian) in Chignin, east of Chambéry, and in Jongieux and Monthoux on the steep western side of the Massif du Chat.

Additionally, the marly limestones of the Lower Cretaceous (Berriasian) provide suitable sites for vineyards in Apremont and Monterminod, on either side of the Chambéry Valley, while those in Chautagne, to the north of Lake Bourget, and in Ayze (Arve Valley) are supported by Oligocene and Miocene molasse. The majority of the vineyards are planted in the weathering products of the Alps. First amongst these are the glacial moraines found throughout the northern sector that encompasses the crus of Marin, Marignan, and Crépy, where the Chasselas excels, and also in Frangy and part of Jongieux and Chignin. Equally significant is the scree or rubble at the foot of the sheer limestone rock face of the Bauges, from the southern suburbs of Chambéry to Fréterive in the Isère Valley. Then there is the fluvial terrace in Ripaille, just a stone's throw from Lake Geneva. The most dramatic and most recent terroir is the Cru Abymes south of Chambéry, at the foot of Mount Granier. It lies on an apron of rubble that landed there when part of the mountain collapsed on the night of November 24, 1248. More than 5,000 people perished in a rockslide that traveled more than seven-and-a-half miles. Water had seeped into the Valanginian* marls, dislodging the whole strata of

hard limestones, another layer of marls, and the overlaying Urgonian* limestones. For centuries the area remained a rocky, chaotic wilderness until it was eventually repopulated and replanted. Today this "brand new" terroir is particularly to the liking of the Jacquère grape that also thrives in the neighboring Apremont terroir.

The Jacquère and Altesse (also called Roussette) are the predominant white grape varieties in Savoie, yielding crisp, light, fruity wines. Red, rosé, and white wines are produced by the communes at the foot of the Bauges Massif, in Arbin, Cruet, and as far as Fréterive. The dominant grape is the Mondeuse, a tannic, typically Savoie red grape variety that yields deep red, tannic wines. Arbin Vin-de-Savoie is a wine for cellaring. Drawing closer to Chambéry, Chigni produces outstanding white wines from the Jacquère or the Roussanne (Chignin-Bergeron has an intense, sometimes slightly sweet bouquet). The Chautagne cru produces intensely fruity, very successful reds from the Gamay Noir. Seyssel, Monthoux, Marestel, and Jongieux use the traditional varieties, Altesse or Roussette, for their elegant, thoroughbred, violet-scented white wines. Plantings of the Chasselas, south of Lake Geneva, yield delicate wines with scents of hawthorn and hazelnuts.

Chignin is one of the leading vineyards of the Cluse de Chambéry, together with Abymes and Apremont. Chignin produces mainly white wines based on the Jacquère (Chignin Vin de Savoie) and to a much lesser extent the Roussanne (Chignin-Bergeron Vin de Savoie). With plantings now covering the whole of the terroir, the steep slopes of the Bauges Massif offer the only room for expansion.

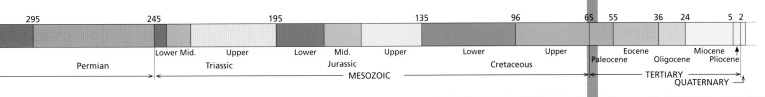

295		245		195		135		96		65	55	36	24	5 2

Lower Mid. Upper   Lower Mid. Upper   Lower Upper   Eocene   Miocene

Permian   Triassic   Jurassic   Cretaceous   Paleocene   Oligocene   Pliocene

MESOZOIC   TERTIARY

QUATERNARY

The Jura

The Jura vineyards (covering 1,830 hectares) line the eastern border of the Saône Plain, facing the Burgundian vineyards of the Côte d'Or. The two areas are similar in some respects but distinct in others. They share the same geological history up to the Miocene, but differ because the Alpine uplift had a very particular influence on the western border of the Jura Massif.

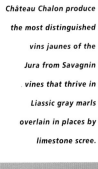

The vineyards of Château Chalon produce the most distinguished vins jaunes of the Jura from Savagnin vines that thrive in Liassic gray marls overlain in places by limestone scree.

The region remained submerged throughout the Mesozoic except for a few periods in the Lower Cretaceous when the seas retreated. The thick deposits they left behind are identical to the sedimentary layers on the Côte d'Or. At the start of the Tertiary, the Jura Massif was above water but still unfolded. In the Oligocene, the Bresse region, future valley of the Saône, collapsed between the Jura and the Côte d'Or forming a long north-south graben*. Debris from the borders accumulated in this trough as it sank. Then, in the Miocene, it became completely filled with fine marly sandy sediments deposited by the seas that invaded from the south via the Rhône Valley.

Regular Folding

Also in the Miocene, the trajectories of the two banks of the Bresse graben began to diverge. As Alpine upthrusting reached a climax, it triggered the folding of the Mesozoic layers covering the Jura. The effect was especially marked in the eastern part of the crescent-shaped Jura region (the Upper Jura), producing regular folds that would later serve as a model in geography classes (anticlines, synclines, transverse valleys, coombs, etc).

In the west, the Mesozoic layers were pushed over the edges of the Bresse graben, lubricated by the Triassic saliferous and gypsiferous clays* directly overlaying the Paleozoic basement. Like a waxed surfboard, the entire package of Mesozoic strata

3 800	2 500	Millions of years		540		500		435	410		360	
Archaen		Proterozoic		Cambrian		Orodovician		Silurian		Devonian		Carbonifer
		PRECAMBRIAN								PALEOZOIC		

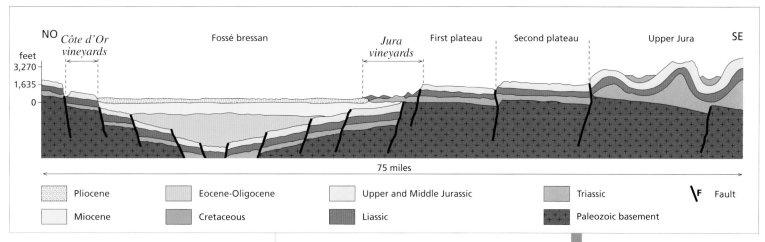

Pliocene	Eocene-Oligocene	Upper and Middle Jurassic	Triassic	↖F Fault
Miocene	Cretaceous	Liassic	Paleozoic basement	

slid westward without folding and started to "surf" over the Miocene deposits of the Bresse graben, covering a distance of three to four miles. With only these softer deposits for support however, the overhanging edge of the plateau cracked and shattered. The hard Bajocian* limestone layers forming the first plateau behind Poligny and Lons-le-Saulnier were faulted and then eroded, exposing the subjacent terrain of the Liassic* and Triassic*.

Triassic and Liassic Supports

The Jura vineyards are established on this fringe of Mesozoic sedimentary strata that surfed over the Miocene deposits of the Bresse graben but lost their protective "carapace" of Bajocian limestone. You can see just how thick these layers are in the village of Salins-les-Bains, built on a salt mine that was excavated out of Triassic strata some 65 feet deep. The Bajocian limestone meanwhile supports the fort of St.-André on the edge of the plateau overlooking the village. The vineyards occupy the intermediary Liassic and Triassic layers. Plantings include local cultivars (the Poulsard, Trousseau, and Savagnin), and the Burgundian varieties Pinot Noir and Chardonnay. The red and rosé wines are honest, fruity, and floral with just the right amount of body to mature after two or three years. The white wines are dry, sometimes with nutty notes of walnuts, hazelnuts, and roasted almonds characteristic of *vins jaunes* (oxidized wines with aromas of green walnuts).

The Multicolored Marls and Clays of Arbois

The Arbois region includes 12 communes centered on the town of Pasteur. It is essentially supported by the Triassic, represented here by marls* and multicolored clays* ranging from green to the color of red wine but always quite dark. The Poulsard excels in these soils, producing rosé-like pale red wines that are low in tannins.

The Gray Marls of Château-Chalon

The other category of soils that is characteristic of Jura vineyards are gray Liassic marls. These support the best terroirs, such as L'Étoile, which takes its name from the star-like cross-section of the fossilized crinoids found in the soil. Gray marls can occur alone when they outcrop on buttes and slopes a short distance from the limestone sill. Those immediately beneath the sill on the other hand are often mixed with talus and limestone fragments from the Bajocian borders of the plateau. A superb example is Château-Chalon, where the exceptional 45-hectare vineyard is dedicated to the production of *vins jaunes*.

The edge of the Bajocian plateau runs in a straight line from Arbois to Miéry, southwest of Poligny, but then starts to weave, jutting into the Liassic in Ménétru-le-Vignoble, then cutting out a deep notch around the blind valleys of Baume-les-Messieurs and Ladoye-sur-Seille. Château-Chalon is located at the exit to this notch. The village perches on a Bajocian-limestone-capped scarp that, due to faulting subsided some 228 feet below the level of the plateau. It is surrounded by vineyards, planted either in talus soils of Bajocian fragments and gray Liassic marls, or in pure gray marls. The highly original Savagnin grape reveals its full potential in these soils, yielding powerful, expressive wines that are the best *vins jaunes* of the region.

Château Chalon, perched on a collapsed block of Bajocian limestone, is a perfect illustration of the disjointed formations that support the Jura vineyards.

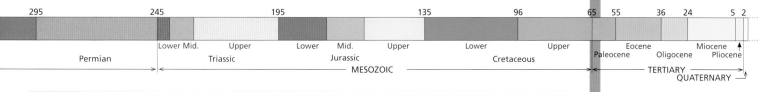

Chile: Vineyards in the Andes

The Chardonnay has found a new home in the sandy arènes of the Casablanca Valley, surrounded by granite hills.

Chile is by no means the biggest South American country, but it does have great viticultural potential. It stretches from the Tropic of Capricorn to the rim of the Antarctic Circle, on a strip of land up to 95 miles wide by more than 2,500 miles long. With strong maritime influences to contend with, the decisive factor in the development of Chile's vineyards is the coastal cordillera.

Geologically speaking, the Andean Cordillera is a recent mountain range, and it is still growing. It was created by the collision of two tectonic plates*, the South American plate to the east, that is slowly drifting westward, and the small, eastward-drifting Nazca plate which is completely submerged below the waters of the southeastern Pacific. When it collided with the South American plate, the Nazca plate began to slide beneath it by a process known as subduction*, slowly sinking into the pool of magma* that lies under the Andean cordillera on the border of the South American plate. At the same time, crustal deformation of the plate's western border produced a succession of landforms which, starting from the

Configuration of the Terroirs of Central Chile.

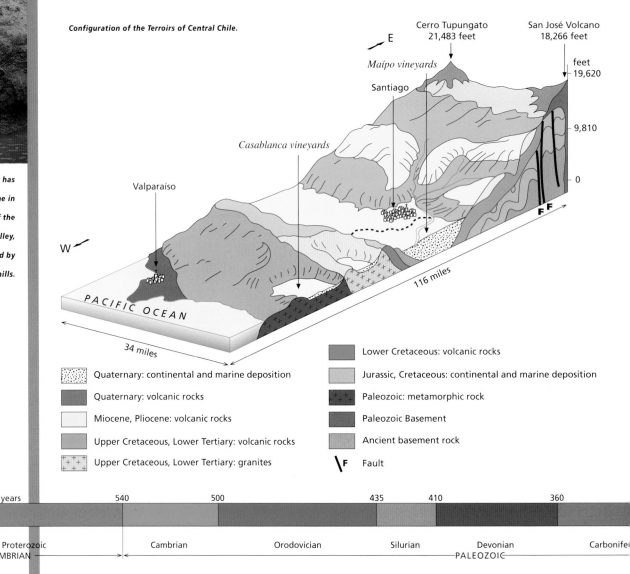

- Quaternary: continental and marine deposition
- Quaternary: volcanic rocks
- Miocene, Pliocene: volcanic rocks
- Upper Cretaceous, Lower Tertiary: volcanic rocks
- Upper Cretaceous, Lower Tertiary: granites
- Lower Cretaceous: volcanic rocks
- Jurassic, Cretaceous: continental and marine deposition
- Paleozoic: metamorphic rock
- Paleozoic Basement
- Ancient basement rock
- Fault

| 3 800 | 2 500 | Millions of years | 540 | 500 | 435 | 410 | 360 |

| Archaen | Proterozoic | Cambrian | Orodovician | Silurian | Devonian | Carbonifer |

PRECAMBRIAN — PALEOZOIC

The Maipo vineyards in Chile's Central Valley region south of Santiago is a perfect example of Chilean viticulture and provides the Cabernet Sauvignon with ideal growing conditions.

Pacific and moving inland, are as follows:

• A coastal cordillera of modest altitude, intersected by rivers flowing down from the main cordillera and largely composed of erosion*- smoothed mountains;

• A collapsed sunken zone of variable width that lies beneath the Pacific at the southern end of the country. This is the foremost wine-producing area.

• The mighty Andean Cordillera that boasts the highest peaks on the American continent. Mount Aconcagua between Valparaíso and Mendoza in Argentina reaches a dizzying 22,759 feet, flanked to left and right by an army of 20,000-foot giants.

While the layers along the borders of the South American continent were being violently folded and uplifted, the continental crust was being pierced by granite intrusions, and the entire cordillera became the focus of intense volcanic activity that upthrust most of the highest peaks in the range.

Between the Humboldt Current and the Cordillera

This overall configuration largely accounts for the Mediterranean-type climate in central Chile, influenced, on the one hand, by the height of the main Andean Cordillera, and on the other by the Humboldt (or Peru) Current from the Antarctic. As it flows from south to north up the coast of Chile, this cold current brings cooler temperatures and mists (*camanchacas*) which are mostly held back by the coastal cordillera.

The central Chilean plain, screened from ocean influences, is intensely luminous. The western part, on the borders of the coastal cordillera, is hot and dry while the eastern margin, at the foot of the main Andean range, is cooler, mainly due to cold air currents blowing down from the Andes, and partly because it is less sheltered from south and southwesterly winds. Hence, the very low night-time temperatures that encourage good color development in red grapes. There are vineyards spread all over this vast region.

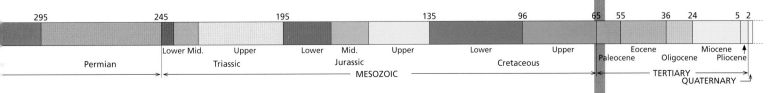

295		245		195		135		96		65	55	36	24	5	2
		Lower	Mid.	Upper	Lower	Mid.	Upper	Lower	Upper		Eocene		Miocene		
Permian				Triassic		Jurassic			Cretaceous	Paleocene		Oligocene		Pliocene	
					MESOZOIC						TERTIARY				
													QUATERNARY		

The Pisco Vineyards

Starting from the north, the first major vineyards specialize in the production of Pisco, an eau-de-vie made principally from Muscat wine. The production zone extends across the basins of the rivers Huasco, Elqui, Limarí, and Choapa. The vines are planted in fluvial alluvia* alongside bodies of water, or in limestone or granite colluvial* deposits from weathering of the surrounding highlands. Plantings in these vineyards, set back from the coast, include the Pisco cultivars, Muscat d'Alexandrie, Torrontel, Muscat d'Autriche, and Pedro Ximénes, together with table-grape varieties.

Valle Central (Chile's Central Valley region)

This vast, 625-mile-long depression*, stretching from Aconcagua to Bío-Bío, is where the coastal cordillera reaches its highest point and provides the most efficient protection from the weather. The geological formations here occur broadly in three parallel bands that correspond to the contour lines.

The coastal cordillera essentially consists of granitic material ranging from the Paleozoic to the Upper Jurassic, Lower Cretaceous, and Tertiary. The Valle Central is a flatland area formed from more or less fine-grained Quaternary fill deposits produced by weathering of the surrounding highlands. These Quaternary deposits extend into the river valleys that cross the coastal cordillera and eventually meet up with the ocean. Tertiary marine sediments are seen in the lower valleys around the mouths of rivers, (such as the Miocene sediments at the mouth of the Rapel). The western flank of the Andean cordillera essentially consists of volcanic material that was deposited at the foot of the Andean volcanoes between the end of the Cretaceous and the start of the Tertiary.

The Valle Central is located between parallels of latitude 32° S and 38° S, and accounts for the bulk of the country's vineyards. The wines produced are labeled with the names of the individual production areas (usually river valleys): Aconcagua, Casablanca, Maípo, Rapel, Curicó, Maúle, Itata, and Bío-Bío. The climate is Mediterranean in type, ranging from semi-arid in the north where irrigation is essential, to more humid in the south. Rainfall increases steadily as you move south, from 12 inches in the Aconcagua valley to 14 inches in Santiago, 28 inches in Curicó, 32 inches in Talca, and 41 inches in Chillán.

The Aconcagua Basin

In terms of pedology, the Aconcagua Basin is essentially composed of alluvial terraces* that range from more or less sandy to stony depending on the area. The best soils are derived from well drained, stony terraces. The Casablanca Valley, on the southwestern side of the Basin, occupies a place apart in Chilean viticulture. For many years, this dry, sandy valley, just below the Atacama Desert and surrounded by granitic massifs, cultivated only the traditional grape variety, the País. More recently, thanks to a cooler climate with strong maritime influences, the Casablanca Valley has successfully introduced the French white grape varieties. The Chardonnay in particular, which now accounts for 80 per cent of the area under vine, is producing very promising results.

The Maípo and Rapel Basins

From the Maípo Basin onward, the coastal cordillera forms a virtually straight edge running south but the width of the Valle Central varies, due to the variable displacement of the cordillera by a network of faults*. The viticultural terroirs of the Maípo and Rapel basins (Cachapoal and Tinguiririca) mainly consist of flat to slightly undulating alluvial terrain that tends to be richer in limestone in the Maípo Valley. The terroirs along the cordillera are a mixture of more or less argillaceous, well-drained, stony colluvial terrain.

South of the Tinguiririca River and in the Talca and Curicó zone, we find stony soils derived from conglomerate* and breccia* and heavier soils originating in volcanic tuff *. From Talca onward, there is a predominance of volcanic ash along the principal Andean range, accompanied by increasing evidence of glacial phenomena (formation of moraines*, lacustrine deposits*).

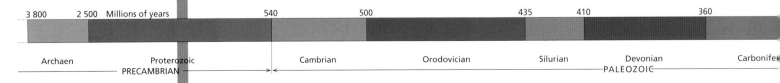

3 800	2 500	Millions of years	540	500	435	410	360

Archaen	Proterozoic	Cambrian	Orodovician	Silurian	Devonian	Carbonife
PRECAMBRIAN				PALEOZOIC		

Table Grapes or Premium Wines?

Vineyard production in the Valle Central varies from one area to another. In the Aconcagua Basin for instance, the emphasis is on table grapes, whereas the Casablanca Valley has recently shifted its focus toward premium white wines based on the Chardonnay. The finest Chilean red wines originate from the Maípo, Rapel, Curicó, and Maíle Valleys where the Merlot and especially the Cabernet Sauvignon produce outstanding results; the Carmenère, a more recent introduction, is also proving a success here. The remaining areas produce either table grapes or table wines (Nuble). In the southern region, where the coastal cordillera loses height, moist air penetrates deep inland. The soils are mainly derived from weathered granite and are usually acidic*, with low fertility and a tendency to poor drainage. In many cases, the terrain has been remodeled by fluvial erosion and shaped into systems of terraces. This is the setting for a number of vineyards, such as those of Mulchén, which have successfully shifted its focus toward white wine production, mainly based on the Sauvignon Blanc.

The coastal cordillera separates the vineyards of the Valle Central from the Pacific coast, screening them from the ocean's moist air.

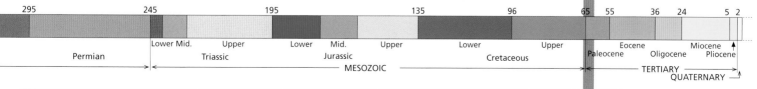

295		245		195			135		96		65	55		36	24		5	2
Permian		Lower	Mid.	Upper	Lower	Mid.	Upper	Lower		Upper		Paleocene	Eocene	Oligocene		Miocene	Pliocene	
			Triassic			Jurassic			Cretaceous									
						MESOZOIC								TERTIARY				
																QUATERNARY		

Argentina:
The Mendoza and Famatina Terroirs

The Argentine vineyards extend for more than 1,250 miles between parallels of latitudes 22° and 40° S (the equivalent zone in the Northern Hemisphere extends from Mauritania to Portugal). The epicenter of the Argentinean wine industry, however, is located around Mendoza, on the eastern flank of the Andean cordillera, almost at the same latitude as Santiago in Chile. This area accounts for more than 90 per cent of output and includes the demarcated areas of Lujàn de Cuyo (the oldest established sub-region) and the two more recently recognized sub-regions of Famatina and Tupungato.

Vines were first brought to Argentina by the Spanish Conquistadors, who probably entered the north of the country via Chile, sometime between 1536 and 1560. But to find out about the characteristic Argentinean wine terroirs, especially those of Mendoza, we need to go east, following the flight path from Buenos Aires to Mendoza. Our route runs more or less straight from west to east, at right angles to the Andean Cordillera. Halfway through the journey, the majestic Andes become a vague presence on the distant horizon, gradually emerging as a majestic, seemingly impenetrable barrier that extends from south to north as far as the eye can see. Highest of all, is the mighty Aconcagua (22,759 feet), flanked by an army of 20,000-foot giants, which we can see looming up behind Mendoza, some 30 miles away as the crow flies.

The Mendoza vineyards, at the foot of the majestic Andean Cordillera, are predominantly planted in the plain, in a region where the sub-desert climate makes irrigation essential.

3 800	2 500	Millions of years		540		500		435	410		360	
Archaen		Proterozoic			Cambrian		Orodovician		Silurian		Devonian	Carbonife
		PRECAMBRIAN								PALEOZOIC		

The vineyards of the Mendoza region are mainly planted in alluvia washed down by rivers from the Andes. The soils are highly varied and range from gravelly to sandy-silty.

An Island of Greenery

The dry, increasingly arid conditions in this region are a far cry from the lush green pampas that was there in the past. Suddenly however, springing out of the desert like an oasis of greenery, are the Mendoza vineyards. These green and pleasant vineyards nestling at the foot of the Andes owe everything to the mountains, especially water – that most precious of all commodities in an area where annual rainfall ranges from a meager 6 to 10 inches. Long before the arrival of the Conquistadors, the American Indians had perfected an ingenious system of water conservation using melted water from the eternal snows. The Spanish then perfected the system, partly for practical reasons (to grow vines and other crops) but also for pleasure, planting trees along the canals

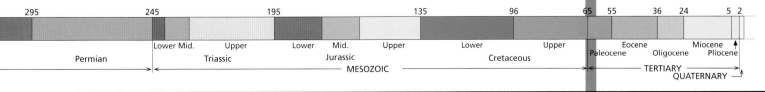

295   245   195   135   96   65   55   36   24   5   2

	Lower	Mid.	Upper	Lower	Mid.	Upper	Lower	Upper		Eocene		Miocene
Permian		Triassic			Jurassic			Cretaceous	Paleocene		Oligocene	Pliocene

MESOZOIC

TERTIARY
QUATERNARY

that gave these waterways a special, leafy charm.

Irrigation is essential in this region, but until recently consisted exclusively of furrow irrigation based on a complex system of canals that distributed the floodwater captured in washes. Because rivers are unreliable water sources, dams and hydraulic pumping equipment were also used to conserve water. This meant that vineyards could only be planted on flat land near rivers.

For some years now, more modern techniques such as groundwater pumping and drop irrigation have brought an end to these constraints and vineyards are increasingly being planted higher up the slopes. One major advantage of low rainfall is that it significantly reduces the need for pesticides since the risk of rot and disease is minimal.

A Terroir Born of the Eroded Andes

Every newborn mountain range is subject to erosion*, and the higher the mountain, the greater the erosion. The Andes were no exception to this rule. Glacial and fluvial erosion stripped away considerable quantities of materials that accumulated first in the valleys, and then in the foothills where the mountains meet the plain. The glacis* formed in the process consists of coarser material close to the mountains, with finer elements farther away. These deposits of Quaternary colluvia* and alluvia* are characteristic of the Argentinean wine terroirs, especially those of Mendoza. They are associated with newly-formed immature soils, that are usually lightweight, deep, more or less sandy or argillaceous, with low alkalinity and low organic content. The vineyards are virtually flat, with no slopes greater than two per cent.

Lújan de Cuyo

Given this combination of vineyards on the plain and deep, fine-grained soils, it is not surprising to find that table wines account for the lion's share of Mendoza output. Most of the vineyards are planted at heights of 1,471-2,616 feet, with two notable exceptions: Lujàn de Cuyo and Tupungato, where the altitude ranges from 2,125-5,232 feet. The vineyards of

Lujàn de Cuyo (2,125-3,270 feet) are watered by the Mendoza River and have a cooler climate and coarser, stonier soils. They produce the finest red wines, making Lujàn de Cuyo the first region to be awarded appellation status in Argentina in 1993.

Tupungato

This more recently developed sub-region, located in the Uco Valley (or Los Harpes Basin) and irrigated by the rivers Tupungato and Tunuyán, is also gaining a reputation for quality wines. The vineyards are located at heights of 2,943-5,232 feet on low-fertility, very stony soils mixed with sand and small quantities of silt. The climate is slightly less dry than around Mendoza, and the terrain consists of a vast alluvial cone* that was formed from eroded material shed from the cordillera. Tupungato is now attracting considerable interest from overseas investors following the 1980s slump in the market.

Famatina

Argentina's Number Two viticultural region, San Juan, occupies similar locations slightly farther north. The terrain is altogether different however in the province of La Rioja, the third most important wine-producing area that mainly extends over the Chilecito Valley. Formed from Quaternary valley-fill deposits from weathering of the surrounding mountains, the valley displays granites* and metamorphic rocks to the east between Chilecito and La Rioja, with Paleozoic and Triassic sediments and granites to the west, in the Famatina Massif. The cooler climate at this higher altitude particularly suits the Torrentés grape, the main regional variety that thrives amid the glittering rocks of the spectacularly colored Famatima Massif. Torrentés yields highly characteristic still or sparkling white wines with a musky bouquet. The Famatima Valley, together with Lujàn de Cuyo and Tupungato, are all demarcated regions.

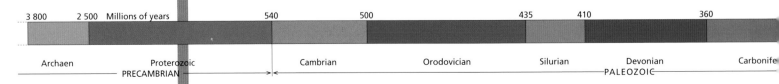

3 800	2 500	Millions of years		540	500		435	410		360
Archaen		Proterozoic		Cambrian	Orodovician		Silurian	Devonian		Carbonife
		PRECAMBRIAN						PALEOZOIC		

Harvesting in Argentina, being in the Southern Hemisphere, starts in March.

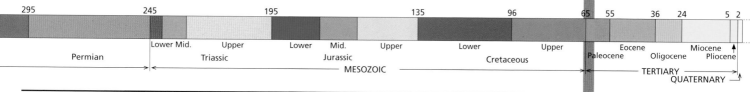

295		245		195		135		96	65	55	36	24	5	2

Permian | Lower Mid. Upper Triassic | Lower Mid. Jurassic Upper | Lower Cretaceous Upper | Paleocene | Eocene Oligocene | Miocene Pliocene

MESOZOIC

TERTIARY
QUATERNARY

California

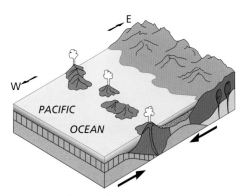

Lower Cretaceous

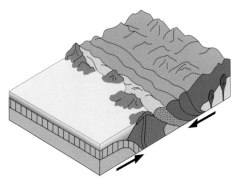

Upper Cretaceous

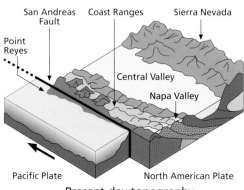

Present-day topography

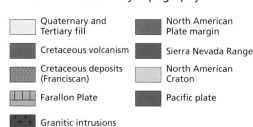

Quaternary and Tertiary fill

Cretaceous volcanism

Cretaceous deposits (Franciscan)

Farallon Plate

Granitic intrusions

North American Plate margin

Sierra Nevada Range

North American Craton

Pacific plate

Because California accounts for nearly 90 per cent of American vineyards, it is often thought to be the only viticultural state in the United States. It does boast the best terroirs on U.S. soil, although the concept of viticultural areas did not exist until 1978. Regulations relating to American Viticultural Areas (AVA's) became mandatory on January 1, 1983. Since then, work has been underway in California to distinguish between particular areas, such as the Napa Valley which currently comprises several AVAs, including St. Helena, Rutherford and Oakville.

For decades, the soil and subsoil were of little importance to Californian producers, who like most New World viticulturalists believed that great wines were largely the product of climate, choice of grapevine, and wine-making techniques. This is surprising when you consider the intense geological interest that this region attracts. California has probably been studied by more geologists, seismologists, and vulcanologists than any other place on earth, and it remains one of the world's most closely scrutinized regions.

The San Andreas Fault

California may be divided into three major units that run parallel to the coast: in the east, the Sierra Nevada (with peaks above 13,000 feet); in the west, the Coast Ranges (never exceeding 7,500 feet); and, sandwiched between these two ranges, the Great Central Valley. All the premier Californian vineyards are located in the mountainous areas of the Sierra Nevada and Coast Ranges, while the Great Central Valley essentially produces bulk wines.

The West Coast is characterized by some particularly complex mountainous formations. California itself lies at the center of an intricate fault system that includes the San Andreas Fault, which cuts through the rocks of the coastal region and causes devastating earthquakes. To understand this unique geological

3 800	2 500	Millions of years		540	500		435	410		360
Archaen		Proterozoic		Cambrian	Orodovician		Silurian	Devonian		Carbonife
		PRECAMBRIAN						PALEOZOIC		

295		245		195		135		96		65	55		36	24		5	2

Lower Mid.   Upper   Lower   Mid.   Upper   Lower   Upper   Eocene   Miocene

Permian   Triassic   Jurassic   Cretaceous   Paleocene   Oligocene   Pliocene

MESOZOIC   TERTIARY

QUATERNARY

phenomenon, we need to go back some tens of millions of years, to the disintegration of the large super-continents. Like the Andean Cordillera, the mountains along the West Coast were created by the collision between two plates, in this case the North American plate (that was drifting westward away from Europe) and the eastward-drifting Farallon Plate (beneath the the Pacific Ocean).

At the end of the Jurassic, the oceanic Farallon plate began to sink beneath the North American plate, by the same process of subduction that continues along the Chilean and Peruvian coasts. A row of islands developed off the western coast of America, separated by an extensional depression with an oceanic crust. In the Portlandian*, the changing configuration of subduction smashed these islands against the coast of California, creating the Sierra Nevada. This process was repeated in the

Cretaceous, starting with the formation of volcanic islands separated from the coast by the Franciscan trench (named after San Francisco). As subduction intensified, blocks of oceanic crust from the basin and material shed by the islands accumulated to form the highly complex structure of the Coast Ranges.

In the mid-Tertiary Period, the Farallon plate was almost entirely consumed beneath the western margin of America. Behind the Farallon plate came the immense, northwestward-drifting Pacific plate – the biggest on Earth – that came into contact with the margin of the North America plate, but this time there was no collision. Instead, the two plates started to slide laterally against each other along the San Andreas Fault. The notorious fault line forms a distinctive gash in the landscape that stretches southward down the coast of California, from the Cape Mendocino viticultural area all the way to the Gulf of California,

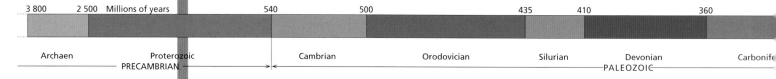

| 3 800 | 2 500 | Millions of years | 540 | 500 | 435 | 410 | 360 |

Archaen | Proterozoic | Cambrian | Orodovician | Silurian | Devonian | Carbonife
PRECAMBRIAN | PALEOZOIC

passing east of Point Reyes, then beneath San Francisco, down the middle of the Santa Clara Valley AVA, along the aptly-named Tremblor Ranges in the Central Valley (away from the coast), and south toward Los Angeles.

Thus all the AVAs south of San Francisco, except for Livermore and San Jose, are located on the Pacific plate, which over the past 20 millions years has been sliding along the North American plate margin at the rate of approximately two inches a year, covering a distance of more than 500 miles.

The Vineyards of the Northern Coast

The cream of California's vineyards are located in the Napa, Sonoma, Mendocino, and Lake counties, north of San Pablo Bay that extends San Francisco Bay in the Coast Ranges. The southeast-northwest running mountains were formed in the Middle Cretaceous and consist of a highly complex structure of blue schists* (of sedimentary or volcanic origin), Jurassic detrital rocks that were metamorphosed* in the course of orogeny, and ophiolites*, from slices of the oceanic crust that were upthrust by compression. In the Miocene, this overall structure was more or less overlain by sediments, while in the Pliocene it underwent intense volcanic activity. At the end of the Tertiary and the Quaternary, sediments accumulated at the base of the valleys, followed by stony alluvial terraces.

The Exemplary Napa Valley

The Napa Valley, which owes its name to a small city north of San Francisco, forms a long tapering ribbon that extends in a northwesterly-southeasterly direction for nearly 50 miles. Its width ranges from five miles at the widest point near the city of Napa to barely one mile where the valley narrows to the north near the town of Calistoga. It is cradled here between the sheer walls of the Palisades at the foot of Mt. St. Helena to the east, and the forested Mayacamas Mountains to the west. From Mt. St. Helena it runs southward towards the San Pablo Bay, eventually reaching Napa. The bulk of the vineyards are located in the more or less flat to slightly undulating terrain formed from Tertiary and Quaternary continental or marine sedimentary fill. The vines are cultivated in monoculture in a wide diversity of soils that, depending on the quality of deposition, range from light sandy-silty soils to very stony alluvial terrain created by rivers. The best soils are found in Oakville and Rutherford. The deeper, heavy argillaceous terrain is less appropriate for quality plantings.

Classification of the Napa Valley AVAs

There are currently 13 AVAs (American Viticultural Areas) in the Napa Valley, including three on the flat valley floor: Saint Helena, Rutherford and Oakville. Stags Leap, the first AVA to be defined, in 1989, on the basis of its distinctive soils, is located south of the rocky escarpment that marks a change in climate and the middle of the valley, north of Youthville. The vineyards are partly established in the argillaceous-silty sediments of the flat zone and partly in volcanic soils derived from Miocene basaltic magma*. Despite the very particular soils in this area, the pattern of plantings is fairly random, with vineyards scattered over the valley sides, supported by a wide variety of different rocks. The Howell Mountain and Atlas Peak AVAs to the east lie on tuff* and volcanic lava, and on rocks of the Franciscan Complex. The Mt. Veeder AVA farther south rests on Cretaceous sandstone* terrain. This AVA separates the Napa Valley from the Sonoma Valley and extends southward as far as the Carneros AVA. North of the road from Napa to Sonoma, the Mt. Veeder vineyards grow in tuff and basaltic lava, while south of the road they rest on Pliocene sands and gravels. Closer to San Pablo Bay however, the soil may be less of an influence than the climate, judging by the European flavors of the Pinot Noir and Chardonnay wines. In the rest of the valley, the Cabernet Sauvignon and the Chardonnay are the leading grape varieties, plus the Chenin Blanc, Gewürztraminer, Riesling, Sauvignon Blanc and Semillion for white wines; and the Cabernet Franc, Merlot, Syrah, Sangiovese, and Zinfandel for red wines.

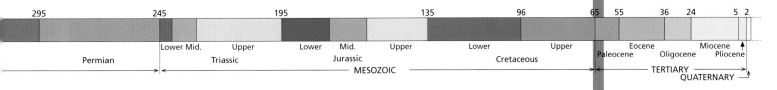

295		245		195		135		96	65	55	36	24	5 2

| Permian | | Lower Mid. | Upper | Lower | Mid. | Upper | Lower | | Upper | | Eocene | | Miocene |
| | | | Triassic | | Jurassic | | | Cretaceous | | Paleocene | | Oligocene | Pliocene |

MESOZOIC — TERTIARY — QUATERNARY

Switzerland:
The Vaud and Valais Cantons

As you can see at a glance from any viticultural map of Switzerland, Swiss vineyards lie mainly along rivers and around lakes. Vines in this continental climate need well-exposed slopes, and on the borders of Lake Geneva they enjoy a double dose of sunshine: from the sun itself and from its reflection in the waters of the lake.

The Cru Dézaley, in the heart of the Lavaux. The Chasselas thrives here in soils derived from pudding stone.

In terms of physical geography, Switzerland could not be simpler. What you have is a lower "plateau" zone of more or less undulating land, bounded to the northwest by the Jura Mountains and to the south and southeast by the Swiss Alps. Its broad geological features are just as obvious.

A Terroir of Moraine and Molasse

For most of the Mesozoic Era, the region was submerged beneath the Alpine Sea. At the start of the Tertiary, the sea retreated and, in the Oligocene relaxation phase, the central part of the country (the "plateau") collapsed between the Jura and the future Alps. The resulting trough was briefly invaded by the sea in the Rupelian, depositing marine sediments. In the Chattian, deposition included not only freshwater molasse from the filling of the trough but also pudding stone, such as that on Mt. Pellerin north of Lake Geneva, which is derived from the dismantling of the nascent Alpine range.

Material shed by the range gradually piled up in front of it as if pushed there by a bulldozer, making the Earth's crust sag. The fill deposits in the graben are consequently much thicker along the borders of the Alps. In the Miocene (Aquitanian, Burdigalian), the waters of the Mediterranean Sea invaded via the Rhône Valley and reached all the way to northern Austria (peri Alpine trough) before retreating. In the Tortonian, the graben became entirely filled in by fresh water molasse. In the Quaternary, the whole region lay under glaciers that helped to dig the famous Swiss lakes of today (notably Lake Geneva), then retreated, leaving vast expanses of moraine* that covered the Tertiary molasse.

The Vaud Canton

The cantons of Vaud and Valais are Switzerland's two foremost wine-growing regions, and they are completely different. The Vaud vineyards are entirely located on the "plateau" and include the Côte region, a long strip of terrain between Geneva and Lausanne, within a few miles of Lake Geneva and parallel to its borders. The plateau features an impressive, often steep-sided south-southeast-facing talus slope of Chattian molasse. Known locally as "molasse grise", it only outcrops in the upper half of the slope, while the top is overlain by moraine, which has spread to most of the lower slope through the action of erosion*. The Chasselas, Pinot Noir, and Gamay Noir thrive in these formations, which continue east of

3 800	2 500	Millions of years	540	500	435	410	360
Archaen	Proterozoic		Cambrian	Orodovician	Silurian	Devonian	Carbonife
	PRECAMBRIAN					PALEOZOIC	

Loire

In the Loire Valley, the 2002 vintage is being highly rated, although crop size is down compared to 2001. The growing season was mostly mild, with little rain and very good weather continued through September. The harvest for Muscadet began on Sept. 16, yielding grapes of high quality.

Sugar and acidity were good throughout the region with little rot. Overall, the crop level was reduced by up to 30%.

Loire 2003 - the monster vintage?
publication date: Dec 30, 2003

It will be fascinating to see the effects of last summer's exceptional heat on the 2003 crop in Europe's cooler wine regions. Vignerons in England are already getting terribly excited about unprecedented levels of ripeness and in France, the Loire is the region most likely to be most dramatically affected by the summer heatwave.

The following is a report from Charles Sydney who, with his wife Philippa, moved from England to the Loire to be a wine broker in 1988. It is he who has already supplied this site with a few interesting images such as that strange insect. He seems hysterically pleased with the vintage but others are suggesting a bit more caution. As elsewhere sugar levels rose much earlier than other components genuinely ripened. There was a high incidence of grape burn. Will the wines turn out to be in balance? Etc, etc.

Those who would like to taste for themselves should take a look at the Angers Wine Fair, 2-4 February 2004 (www.salondesvinsdeloire.com) It is a great showcase for this undervalued region. I'd love to go but I really must stay at my desk alas.

2003 Loire Valley Vintage Report by Charles and Philippa Sydney
Wow... Certainly the earliest harvest in the Loire since the quasi-mythic 1893, 2003 is quite some vintage, breaking all records for ripeness and precocity...

 But is it good? Is there enough acidity?

The precipitous slopes of the Sion vineyards (in the Valais) on the rightbank of the Rhône, home of the famous Swiss wines, Fendant and Dôle.

Lausanne, on the Côte de Lavaux, where the vineyard borders directly on Lake Geneva. Closer to the Alps, the molasse in enriched by coarser elements from weathering of the mountains. The Pellerin pudding stone supports the vineyards in certain parts of Dézaley. Home of the greatest white wines in Switzerland, the Dézaley vineyards are the kingdom of the Chasselas that flourishes on precipitous, terraced slopes lined with small earth-retaining walls. Dézaley Chasselas has an instantly recognizable mineral bouquet, with notes of lime blossom and a touch of bitterness at the finish.

The landscape looks completely different farther up the Rhône River, at the eastern tip of Lake Geneva, where the Alps commence. In this mountainous Chablais region, where the vines find a few well-sheltered niches here and there, most notably on Triassic argillaceous stony terrain.

The Valais Canton

From its source in southern Switzerland, the River Rhône flows to Lake Geneva, then through a narrow valley that runs northeast-southeast to Martigny. The dog-leg in the river at this point marks the beginning of the Valais, a canton with a remarkably warm, dry climate (less than 25 inches annual rainfall) thanks to the föhn wind and the protective screen of the Italian Alps.

Sheltered from the weather, the vineyards extend upwards on the rightbank of the Rhône, to heights of more than 3,270 feet. There are also a few vines on the leftbank of the river (in Ridde, for instance). Most of the plantings are on precipitous slopes of essentially Jurassic terrain, consisting of schists and limestones that were metamorphosed under the intense pressure generated by the Alpine uplift. The Valais is famous for its Fendant, a superb dry white wine made from the Chasselas, and Dôle, a red wine blended from the Pinot Noir and Gamay. In some years, dessert wines are also produced from late-harvested, nobly rotted Sylvaner (known locally as the Johannisberg), Marsanne, and Pinot Gris.

The soils are predominantly derived from limestones (Cretaceous) between Sierre and Leuk, while those at the other end of the Saillon Valley in Fully and Martigny feature outcrops of basement granite that produce superb results in the Gamay. Additionally, there are several alluvial fans* in the Valais, the largest being in Chamoson, famous for is wines based on the Pinot Noir, Gamay Noir, Sylvaner, Chasselas, and local varieties such Humagne Rouge and Humagne Blanche, Petite Arvine, and even the Cornalin. The soils in this area – a mixture of stones shed by weathering of the peaks, plus small amounts of clay and silt – have been planted with vines for centuries. Glacial meltwater keeps the fan moist through capillary action, preventing the soils from drying out in summer.

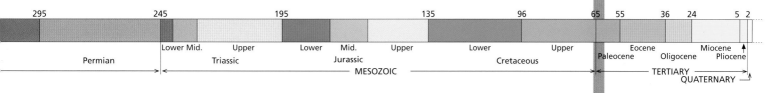

295		245		195			135		96	65	55	36	24	5	2

Permian	Lower Mid. Upper	Lower Mid.	Upper	Lower	Upper	Paleocene	Eocene	Oligocene	Miocene Pliocene
	Triassic	Jurassic		Cretaceous					

MESOZOIC

TERTIARY

QUATERNARY

The Chasselas

The Chasselas is the only grapevine to rank as an AOC table grape (Chasselas de Moissac from Tarn-et-Garonne) and wine grape (the AOC wines of Vin-de-Savoie, Alsace, and Pouilly-sur-Loire are all made from the Chasselas). It has also made the reputation of Swiss wines, most especially those from the Vaud canton, where the Chasselas displays unique characteristics.

Chasselas wine grapes are to Switzerland what the Riesling is to Germany. The Chasselas is most at home in Switzerland, where it is the number one variety, beating the Pinot Noir (in second position) and the second most common white grape variety, the Müller-Thurgau (in fifth position). French plantings of the Chasselas in the Haute-Savoie and Alsace are actually the southern and northern extensions of Swiss plantings. Switzerland has 14,900 hectares of vines, producing 1.2-1.3 million hectolitres of wine per year, of which 58 per cent are white wines and 42 per cent are red wines. The Chasselas accounts for 5,490 hectares of the total Swiss area under vine, compared with 4,560 hectares for the Pinot Noir and just 698 hectares for the Müller-Thurgau.

The Chasselas in the Vaud Canton

This canton that extends from Lake Geneva to Lake Neuchâtel on the borders of the French Jura region is almost exclusively planted to the Chasselas. It accounts for 2,700 of the 2,782 hectares planted to white grape varieties (97 per cent) and for 70 per cent of plantings overall (red and white grapes).

On the northern banks of Lake Geneva, in particular, we find the Chasselas planted throughout the Côte region, north of Nyon, and especially on the precipitous Oligocene slopes of Lavaux between Lausanne and Montreux. It performs especially well in the Pellerin pudding stone of the Cru de Dézaley, yielding beautifully balanced wines with mineral and floral aromas (lime blossom) that rank among the most expensive

of all Swiss wines. The Chasselas also excels in the Chablais region, where soils derived from strongly folded Mesozoic rocks, ranging in age from the Triassic to the Cretaceous, mark the commencement of the Alpine formations.

The Chasselas in the Valais Canton

The Chasselas accounts for 1,740 hectares of the 2,296 hectares of white plantings in the Valais, where the geographical situation is quite different, although the slopes are just as steep. The vineyards are planted in the heart of the Alps, along the valley of the Rhône River, where it flows northwest to southwest from Brig to Martigny. The Chasselas is mainly established on the sunnier, right bank of the river and the downhill stretch of the left bank. The grape produces the celebrated Swiss Fendant, a crisp, fresh wine with a mineral bouquet.

The vineyards in this area are supported by a variety of formations, ranging from Mesozoic limestone* formations upstream to *schistes lustrés** around Sion and granites* in the Martigny region (where the Gamay dominates). There are also well-developed vineyards in the alluvial fans* that have formed where the Alpine valleys meet the Rhône Valley; the largest alluvial fan is in Chamoson.

There are more Chasselas plantings north of Geneva, on the borders of Lake Geneva, and around Lakes Neuchâtel and Bienne where Miocene molasse* overlaid with a thin veneer of moraine* meets the first Cretaceous limestone slopes of the Jura.

The Chasselas in France

On the southern bank of Lake Geneva in the Haute-Savoie region, the Chasselas occupies Tertiary and Quaternary formations that are the southwestern extensions of the formations in the cantons of Vaud and Geneva. A stony terrace on the borders of the lake is home to some 20 hectares of Chasselas plantings on the Ripaille estate, while Würm moraine of similar acreage supports the Cru Marin slightly farther south. A few miles south, we come to Crépy, where argillaceous morainic deposits known as boulder clays support the most

extensive Chasselas plantings in the Haute-Savoie (around 100 hectares). These formations continue in the neighboring Cru Marignan (five hectares), with slightly more Miocene molasse at the foot of Mt. Boisy. The local Chasselas wines are lighter than those from the Swiss side of Lake Geneva, with a crisp, fresh quality and aromas of hawthorn and hazelnuts.

The Chasselas also accounts for a few plantings in Pouilly-sur-Loire, miles away from its beloved Alps. It particularly likes the flint-rich Eocene siliceous terrain (St.-Andelain) where it yields soft, light wines smelling of citrus and white flowers.

The Chasselas planted here at Mont-sur-Rolle is the predominant grape variety in the Côte de Lavaux, north-west of Lake Geneva.

Corsica

This island mountain features a variety of indigenous cultivars that are perfectly adapted to this craggy, sunny environment with sea on all sides. The island may be divided into three broad sections, with a granitic part in the south and west and schistose soils in the northeast, separated by a narrow band of limestones. A range of vineyard sites all produce wines that clearly express their maritime origins.

The indented coastline of the Patrimonio region opens onto the Gulf of St.-Florent.

Corsica and Sardinia, both with a tradition of wine-growing, sit side by side in the middle of the Mediterranean Sea. Now closer to Italy than France, until the mid-Tertiary, the two islands originally formed part of the Pyrenean-Provençal axis and lay a few hundred miles farther to the northwest. In the Miocene, the whole area commenced a rotation of approximately 30° southeast around a point that would now lie in the vicinity of La Spezia, on the Italian coast.

Ancient Corsica and Alpine Corsica

Corsica's geological structure and history is the same as that of the Provençal axis. The western and southern two thirds of the island, *Ancient Corsica*, is composed of different types of granite and sits on a slice of the ancient Hercynian range. It is bordered to the east by a line running from the west of St.-Florent to Corte and as far as Solenzara on the eastern coast.

The rest of the island, *Alpine Corsica*, extends all the way to the Cap Corse and consists of *schistes lustrés*. These are marine sediments that were deposited in the Mesozoic in very deep seas and thrust against the Hercynian basement in the course of the Alpine uplift. They are mixed with green rocks (ophiolites) that originated in submarine basalt lava flows. In the Miocene, most of Alpine Corsica was submerged by seas and overlain by sediments. Subsequent folding and deformation largely wiped out these Miocene deposits that now only survive around Bonifacio, St.-Florent, and especially on the eastern coastal plain from Cervione to Ghisonaccia. In the Quaternary, this plain was overlain with outwash by rivers flowing down from the Corsican mountains.

Coastal Vineyards

Vines in this Mediterranean island par excellence have always been planted along the coast – although with mixed fortunes. The eastern plain changed face in the 1960s when it was widely replanted by French colonialists returning from North Africa. Twenty years later, these hills were almost entirely covered with grapevines. More recent years have brought a

3 800	2 500	Millions of years		540		500		435	410		360	

Archaen	Proterozoic	Cambrian	Orodovician	Silurian	Devonian	Carbonifer

PRECAMBRIAN — PALEOZOIC

new downturn in wine-growing here, despite the excellent soils on stony terraces.

The Vin-de-Corse and Vin d'Ajaccio AOCs: Granitic Terroirs

Traditional Corsican plantings took a beating in the phylloxera epidemic and are also in decline. Vineyards extend along much of the Ancient Corsican coastline, from Porto-Vecchio to Sartène, Ajaccio, and Calvi, exclusively planted in granitic soils. The best vineyards are in the Ajaccio area that is an appellation in its own right. They line the bays of the serrated coastline, surrounded by lush vegetation on well-drained, granitic slopes with ideal sunshine. The sandy arenaceous soils here suit the Sciaccarellu, the main grape variety, yielding mainly red and rosé wines with elegant body, violet-scented aromas, and sometimes a slight whiff of the sea.

Patrimonio: Between Schists and Limestones

Vines also line the coastline of Alpine Corsica, but on two quite different types of terrain. The Patrimonio appellation area sits on the eastern border of the Gulf of St.-Florent in the Nebbio district, named after the fogs that frequently blanket the region in the fall. This part of the island corresponds to a thrust zone of sedimentary layers that border into *schistes lustrés*; these form the peaks that guard the entrance to the Teghime Pass coming from Bastia. The vineyards mainly occupy a small valley extending from Oletta to Patrimonio, on Paleozoic to Cretaceous geological formations that run north-south, becoming increasingly recent as you move west toward St.-Florent and the Mediterranean. On the valley floor, a row of limestone spurs juts out of the earth like standing stones. To the west, the valley is bounded by sheer cliffs composed of the Miocene molasse that caps the vine-covered Eocene slopes.

There are vineyards on most of these geological formations, producing mainly red and rosé wines from the Niellucciu, which flourishes in the argillaceous-calcareous soils. This type of soil also favors the white grape variety Vermentinu. Niellucciu reds

are packed with character, combining good body with spicy aromas overlaid by notes of raspberry and black currant. The white wines are rich, silky and remarkably aromatic with scents of white flowers and exotic fruits.

Muscat-du-Cap-Corse, the third and final Corsican AOC, lies on the *schistes lustrés* and green rocks that form the bulk of the terrain at the northern tip of the island. The vineyards here produce a fresh, elegant Muscat wine made entirely from the Muscat Blanc à Petits Grains.

Above: the vine-covered Eocene slopes of the Oletta Valley in the Patrimonio area lie at the foot of the rocky Miocene barrier that separates the valley from the sea. Scattered amidst the garrigue of the valley floor are needles of Jurassic rock.

Left: the vineyards on the granitic slopes of Ajaccio enjoy various exposures.

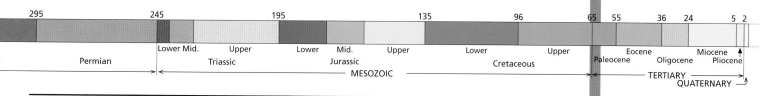

Volcanic Terroirs

Volcanoes erupt when magma (a mixture of hot molten rock, gases, and other materials) wells up beneath the Earth's crust and is ejected at the surface by an intense displacement of the crust. From the distribution pattern of the thousands of volcanoes that remain active today, we know that they mainly occur in rift zones and collision zones. Rift zones are caused by crustal separation and spreading, mainly at the mid-oceanic ridges (for example, the mid-Atlantic Ridge). Collision zones are where two plates are moving towards each other.

Volcanism also occurs along grabens where the fault lines on each side create easy escape routes for the magma. The mineral composition of the materials ejected at the surface varies significantly, from very friable and likely to crumble within a few months (it is the Peléean type, named after Pelée Mountain in the Martinique) to tough and highly resistant to weathering (for example, basaltic magma).

Many of the world's vineyards lie on volcanic terrain. In Greece, for example, especially in the Aegean Islands and most famously on the island of Santorini. Volcanic terroirs also support the Sicilian vineyards of southern Italy. The inhabitants of Pompei, before they were buried beneath volcanic ash and lava in 79 AD, originally planted their vineyards on the fertile slopes Vesuvius. The northwestern United States lies on vast expanses of basaltic formations that support the vineyards of Oregon, where the Pinot Noir performs so well.

In France, on the other hand, where each orogenic episode marked a period of intense volcanic activity, very few vineyards, in fact, lie on volcanic soils. The only exceptions are a few plantings in southern Alsace, the Beaujolais region (pre-Hercynian or Hercynian Orogeny), and the Côte du Forez. In the volcanic mountains of the Auvergne region, formed in the course of the Alpine uplift and active until very recently, the vineyard terrain is protected by volcanic formations, but the vines are not rooted in the formations themselves.

The vineyards of the Geria Valley on the island of Lanzarote (Canary Islands) are carved out of volcanic ash.

The Vineyards of the Côtes d'Auvergne

The favorable vineyard sites of the Auvergne area are only partly the product of the intense volcanic activity that led to the formation of the region's famous volcanoes in the Miocene. The origins of the terroirs may be volcanic, and sheltered by volcanic mountains (the *puys*), but the growing medium is essentially composed of argillaceous limestones mixed with volcanic debris. The slopes that border the Limagne area are carpeted with vineyards, extending up the mountainsides to heights of 1,144-1,635 feet.

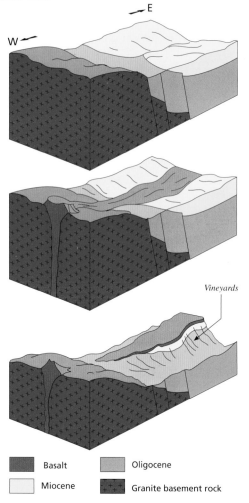

◼ Basalt	◼ Oligocene	
◻ Miocene	◼ Granite basement rock	

The Massif Central covers a 30,888-square mile-area, which is nearly one-seventh the area of France. Its impressive series of volcanic ranges received their final facelift in the Alpine Orogeny of the Tertiary Period, marking the end of three major events that shook the Massif Central from the mid-Tertiary to the end of the Quaternary.

Formation of the Massif Central

The first event in the Oligocene, coinciding with the creation of the Bresse and Alsace grabens, was the collapse of the Puy-en-Velay graben* and especially the Limagne graben (Allier Valley), along a network of north-south-facing faults*. These depressions then received Oligocene and Miocene deposition of marls and limestones, followed by lacustrine limestones). The second event, which had a significant impact on the viticultural slopes of the Auvergne, was the intense volcanic activity that started with basalt lava flows in the Miocene and gathered momentum in the Pliocene, producing the major volcanoes of Plomb du Cantal and Mont-Dore. More tongues of lava spilled out in the early Quaternary, forming the plateaus of Aubrac and Coirons. The Puy Range, location of the celebrated Puy de Dôme, west of Clermont-Ferrand, is a recent formation; the last eruptions in the Auvergne took place less than 6,000 years ago, when the region was already inhabited by people. The hills and peaks we see today took shape as the alpine uplift reached a climax. This was the third and final event, involving intense upthrusting of the basement rocks of the Massif Central, which started toward the end of the Tertiary and continued into the Quaternary. In the process, rivers gouged deep channels through the hillsides.

Inverted Relief

The vineyards of the Côtes d'Auvergne lie on either side of the town of Clermont-Ferrand, scattered along the western border of the Limagne graben across a distance of nearly 50 miles. At a glance, there appears to be no difference between this location and that of neighboring vineyards such as the Côte Roannaise and the Beaujolais. In fact, however, the

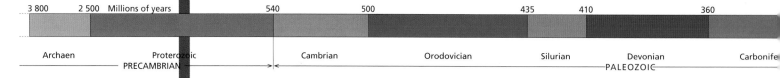

3 800	2 500	Millions of years		540	500		435	410		360

| Archaen | | Proterozoic | | Cambrian | Orodovician | Silurian | | Devonian | | Carbonife |
| | PRECAMBRIAN | | | | | | PALEOZOIC | | |

220

The sloping vineyards of Chanturgue, overlooking the town of Clermont-Ferrand, are planted in marly limestone layers overlain by feldspathic sands. Basalt lava formed the cap rock of the plateau in the Miocene, shedding basalt fragments that mixed with marly limestone colluvia to form the soil.

Auvergne area stands apart on account of its volcanic origins. But we need to go back to the Miocene to gain a proper understanding of the processes at work. It was at this time that the Limagne graben became filled with more than 6,000 feet of sands, arkose*, Oligocene lagoonal and lacustrine limestones, and Miocene sands. Volcanoes developed right across the Hercynian basement* and along the network of faults separating it from the graben, spilling out tongues of basalt lava into the valley floors that flowed toward the Limagne and smothered the basement and the Oligocene sediments.

Basalt is a tough, heavy rock, so erosion mainly affected the surrounding Oligocene sediments, leading to the development of an inverted relief. Lava formed the cap rock of the plateaus (hills of Chateaugay, Chanturgue, Corent, and Boudes) on which the vineyards are now established, mainly planted in Oligocene terrain that is mixed with basalt colluvia* stripped from the plateau, but never in the basalt itself.

The Unusual Terroirs of Chateaugay and Corent

These privileged, sheltered locations, screened from north winds and cold air currents from the Massif Central, are well suited to the Gamay, which is blended with small quantities of the Pinot Noir to produce fruity, quaffable red wines. In some terroirs, such as Chateaugay and Corent, the Gamay yields wines with very particular characteristics. The slopes of Chateaugay feature peperites*, unusual volcanic deposits created when basalt lava is invaded or intruded by wet sediments. Sudden cooling and fragmentation created glassy granules that together with the Oligocene sediments now form a well-drained terroir that is quick to warm up and is ideal for the Gamay. The vineyards of Corent are planted on the volcanic slopes of a cone that surged through the Oligocene sediments and resisted erosion; they are famous for their pale pink, richly aromatic rosé wine made from the Gamay.

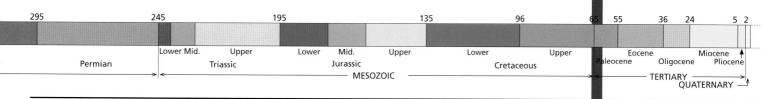

Hungary: From Tokaj to Lake Balaton

ungary is divided into two quite different sections by the Danube River. To the west, the hilly region of Transdanubia centers on Lake Balaton, while to the east lie the sandy expanses of the Great Hungarian Plain, bordered to the north by the foothills of the Carpathians. At the northeastern tip of Hungary are the celebrated vineyards of Tokaj-Hegyalja, where unusual soils, a warm, humid microclimate and generations of wine-growing expertise combine to produce a legendary wine.

The southern tip of the Zemplén Mountains and the Tokaj vineyards.

The volcanic activity that led to the development of some of Hungary's viticultural soils was closely linked to the Alpine Orogeny*. The French and Swiss Alps extend eastward into the Austrian Alps, followed by the Tatras Mountains in Slovakia, and then the giant arc of the Carpathians. The Carpathians envelope northern, eastern, and southeastern Hungary in their mighty embrace before continuing into the Balkans (Bulgaria).

The Carpathian Mountains

By the end of the Jurassic, eastern Europe was feeling the first effects of the collision between the European and African continents. This was long before the Alpine uplift started in France in the Tertiary; yet the Carpathians and the Alps are the product of the same orogenic process. They were both pushed up when the tip of the northward migrating African

The cellars of the Princes Rakoczi in the village of Tokaj are dug out of Mount Tokaj dacite rock.

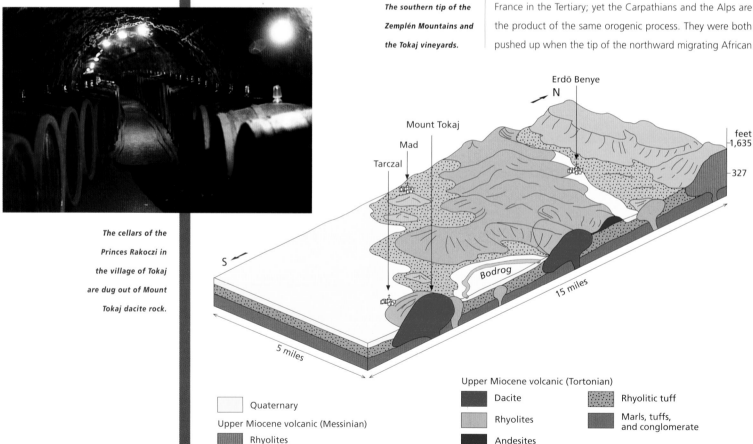

| Quaternary |
| Upper Miocene volcanic (Messinian) |
| Rhyolites |

Upper Miocene volcanic (Tortonian)

Dacite	Rhyolitic tuff
Rhyolites	Marls, tuffs, and conglomerate
Andesites	

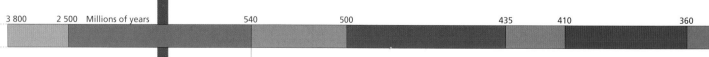

| 3 800 | 2 500 | Millions of years | 540 | 500 | 435 | 410 | 360 |

| Archaen | Proterozoic | Cambrian | Orodovician | Silurian | Devonian | Carbonifer |
| PRECAMBRIAN | | | | | PALEOZOIC | |

plate ploughed into southern Europe beneath the Mediterranean Sea, between Italy and Greece. The climax of the Alpine uplift in the Miocene was associated with the first episode of intense volcanism. Known as andesitic* volcanism, it was characterized by eruptions of thick, highly viscous material that spewed out within the arc formed by the Carpathians (forming the Volcanic Carpathians), from the center of Slovakia (forming the Tatras Mountains) to the Matra and Tokaj regions, and all the way to Transylvania.

As the mountains swung in a wide, crescent-shaped arc extending northward and eastward, the Earth's crust was stretched* and started to sag, forming the Pannonian Basin that now corresponds to the southern core of Hungary. The sea occupied this basin for a while, followed at the end of the Miocene by the immense Pannonian Lake that plastered the entire region with thick deposits of sands, sandstone, and argillaceous marls.

In the Pliocene, this lake dried up and a new volcanic period started. These eruptions were of a basaltic* type, characterized by highly fluid lava flows that engulfed the Pannonian lacustrine sediments. In the course of the Quaternary, these were gouged out by erosion*, which gnawed away at the Carpathian Range and its volcanic deposits. Wind-blown deposits, meanwhile, accumulated throughout the region.

Tokaj in the Zemplén Mountains

The vineyards of Tokaj occupy the southeastern and southwestern slopes of the Zemplén Mountains, a small, north-south-facing volcanic range within the Carpathians. The range begins with Mount Tokaj itself, its 1,713-foot peak towering over the Great Hungarian Plain. The range extends toward the Slovakian-Hungarian border, reaching heights of 2,929 feet on Mount Nagy-Milic where it merges with the main Carpathian range. The Zempléns are formed from volcanic rocks and are essentially andesitic* (the lava ejected as almost-solidified rocks, depending on the viscosity of the magma), trachytic* (marking the flow of more viscous lava), and rhyolitic* (the volcanic equivalent of granite, forming the bulk

Above:
The vineyard slopes
of Mount Tokaj.

Opposite:
The Disnökö vineyards
are planted in
rhyolitic tuff.

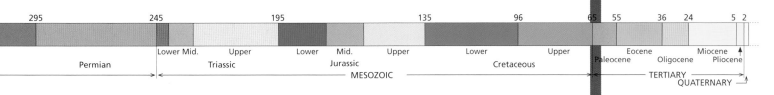

295		245			195			135		96		65	55	36	24	5	2
Permian		Lower	Mid.	Upper	Lower	Mid.	Upper		Lower		Upper	Paleocene	Eocene	Oligocene	Miocene	Pliocene	
			Triassic			Jurassic				Cretaceous							
						MESOZOIC							TERTIARY				
														QUATERNARY			

of the Tatras and Transylvanian Mountains). These eruptions were associated with peripheral accumulations of breccia* (fragments of cemented rocks) and tuff* (layers of falling pumice shards that welded together to form vast, thick aprons). Most of the region's wine cellars are dug out of this rhyolitic tuff, except for those in Tokaj and Tarczal. The Mad cellars are particularly spectacular, forming a labyrinth of underground tunnels that exit at various points throughout the village. The area's clay-type soils are derived from weathering of volcanic rocks at the surface; they yield the finest, most delicately aromatic, and long-lived wines.

Of significantly less viticultural potential are the soils derived from the sandy-argillaceous sediments deposited by Lake

The vineyards of the imperial Hetzolo estate line the southwestern flanks of Mount Tokaj, planted in a mixture of loess and volcanic debris.

Pannonian following the first period of volcanism. The Quaternary saw loess* deposition, most notably on Mount Tokaj and along the valleys of the rivers Bodrog and Hernad.

Between the Rivers Bodrog and Hernad: The Most Famous Dessert Wines of All

These two rivers, particularly the Bodrog, are a large part of what makes the Tokaj terroir so successful today. The Bodrog flows into Tokaj along the southeastern flanks of volcanic hills that extend from Satoraljaujhely, bringing with it the autumn humidity that promotes the misty mornings and sunny afternoons so favorable to ripening and sugar concentration *sur souche* (on the vine rather than after the harvest).

The Tokaj vineyards are planted to the Furmint and Harslevelu, plus small amounts of Muscat. They produce the full gamut of wines, from the very dry to Tokaj Aszu, the great dessert wines that have made the region's reputation. The secret of winemakers' success is the special harvesting technique based on the picking of shriveled botrytized grapes. These are then added to the fermenting musts, producing superb dessert wines capable of aging for more than a century in the tuff caves of the region.

The Eger Viticultural Region

Hungary's second most famous wine-producing area, the Eger vineyards, lie beyond the industrial center of Miskolc, on the southern flanks of the Bükk Mountains. These are the westward extensions of the Zemplén Mountains and reach a height of 3,146 feet on Mount Istállósko. The Bükkalja vineyards are also planted on the southern slopes of the Bükk.

North of the town of Eger, the vineyards are supported by Mesozoic formations that experienced the full force of the Alpine upheaval. Otherwise, the vines grow in deposits of Miocene rhyolitic tuff that can be very thick in places, producing the richest, most elegant wines of the appellation. A superb example is Bikáver d'Eger (Bull's Blood), made from the Kadarka or the Blaufrankisch, blended with the Cabernet varieties and the Merlot. The region also produces remarkably crisp and aromatic white wines from the Welschriesling, Leanika, Traminer,

3 800	2 500	Millions of years		540	500		435	410		360
	Archaen	Proterozoic		Cambrian	Orodovician		Silurian	Devonian		Carbonifer
		PRECAMBRIAN						PALEOZOIC		

and Muscat Ottonel. As previously mentioned, practically all of the region's wine cellars are dug out of rhyolitic tuff.

Somlo: Vines on an Extinct Volcano

The vineyards north of Lake Balaton are characterized by a very different, more recent type of volcanism (Quaternary). A prime example is the highly original Somlo terroir, on the southern fringe of the Little Hungarian Plain, bordered by the hills of the Badácsony region. Somlo is the smallest winery in Hungary (less than 500 hectares), though with a big reputation, entirely planted on the slopes of a hill that is in fact an extinct volcano (called the Somlo). By the end of the Miocene, the region was covered in thick layers of sandstone and more or less argillaceous marls deposited by the immense Pannonian Lake. After it dried up, the lake gradually filled up with stony fluvial deposits. Then in the Pliocene, a large volcano developed on the site of what is now the Somlo hill, pouring basalt lava all over the lacustrine and fluvial deposits. In the course of the Quaternary Period, the only lacustrine deposits to survive erosion were those protected by a carapace of basalt that forms the characteristic rock relief you can see today. The volcano also became deeply eroded, leaving an exhumed plug that now displays its columnar basalt structure. The best wine-growing soils in Somlo

are found on the steepest, highest slopes at the foot of the basaltic covering (tuff and basaltic debris). The area is dedicated to white wine production, based once again on the Furmint, blended with the Harslevelü or the Welschriesling, or with the highly distinctive Juhfark, which yields dry, intensely aromatic wines. Somlo wines are powerful with excellent body, and need to age for several years. According to popular legend, these wines were once reserved for royal weddings.

On the Banks of Lake Balaton

The same basaltic origins characterize the Badácsony region, flanked to left and right by the hills that run along the northern edge of Lake Balaton. Formed from basalt-capped sedimentary lacustrine deposits, the slopes overlooking the lake have been terraced and planted with vines that thrive in a mixture of argillaceous soil and basaltic scree.

The wines from this area range from powerful to soft and elegant, with the best coming from the Badáscony region. In some years, given the right weather conditions, this region produces Aszü wines from botrytized grapes; this is also the only area to cultivate the Keknyelu variety. The northern section of Lake Balaton produces almost exclusively white wines from the Olazriesling, Welschriesling, Riesling-Sylvaner (Rizlingzilvani), Traminer, and Muscat Ottonel.

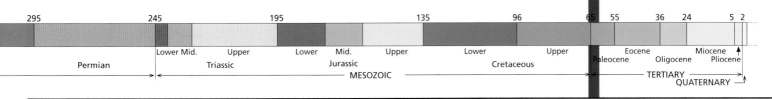

The Canary Islands

The Canary Island archipelago developed in a recent geological epoch less than 20 million years ago. It is located off the African coast, facing Morocco, and consists of 13 islands, of which seven are populated. Administratively part of Spain, the seven main islands are, from east to west: Lanzarote, Fuerteventura, Gran Canaria, Tenerife, Gomera, La Palma, and El Hierro. All of these islands grow vines, especially Tenerife (the number one grape producer with 7,630 hectares of plantings) followed by Lanzarote (2,290 hectares) and La Palma (1,570 hectares). Fuerteventura on the other hand, has less than 20 hectares of plantings.

Vines planted in holes scooped out of volcanic ash are characteristic of the Geria Valley on the island of Lanzarote.

Cross-section of the Canary Island Archipelago.

All of the Canary Islands are volcanic in origin, created by magma that escaped along major cracks in the Earth's crust*, forming cones that more or less project above the waters of the Atlantic Ocean. They range in height from 12,157 feet on Tenerife, where Mount Teide receives a blanket of snow in winter, to 2,253 feet on Fuerteventura and just 1,946 feet on Lanzarote. The remaining islands average heights of 4,905-8,175 feet. The most extraordinary island in terms of viticulture and landscape is Lanzarote, where the vines are planted in holes in the volcanic rock, with a hostile natural environment on all sides. The degree of human effort to cultivate there is nothing short of heroic.

Lanzarote

The oldest island in the archipelago, Lanzarote is also the closest to the African coastline (around 60 miles) and within reach of the scorching Saharan winds. Trees are rare on this virtually desert island, which remains at the mercy of active volcanoes and witnessed its last major eruption in 1824, when Tinguatón, near Tinjajo, spewed lava all over the island.

The effects, however, were not as apocalyptic as those of the eruptions that rocked the island in the previous century, from 1730-1736. Around a dozen villages were buried in volcanic ash*, which also smothered the richest agricultural land under deposits that range from more than six feet deep in the center of the island to 7-11 inches in the north. Far more significant, however, was the formation of some hundred or so craters representing an increase in landmass of more than 77 square miles, which is equivalent to one-fifth of the island's overall area. Today, they form the lunar-like landscapes of the Timanfaya National Park, bathed in magnificent shades of red, ochre, brown, and green.

Heroic viticulture

What happened next ranks as a triumph of human ingenuity over a hostile environment. Wine growers excavated the soil buried beneath the ashes, digging large, inverted-cone-shaped holes (on average three hundred holes per hectare) then planting each one with up to three vines. Small, crescent-shaped dry-

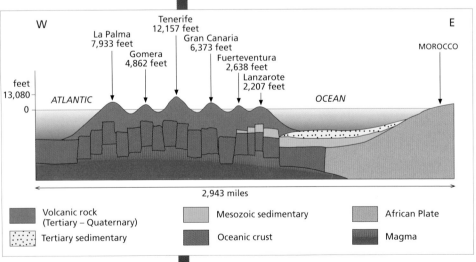

W | La Palma 7,933 feet | Gomera 4,862 feet | Tenerife 12,157 feet | Gran Canaria 6,373 feet | Fuerteventura 2,638 feet | Lanzarote 2,207 feet | MOROCCO | E

feet 13,080

ATLANTIC

0

OCEAN

2,943 miles

- ▇ Volcanic rock (Tertiary – Quaternary)
- ⣿ Tertiary sedimentary
- ▇ Mesozoic sedimentary
- ▇ Oceanic crust
- ▇ African Plate
- ▇ Magma

Timanfaya National Park: an apocalyptic landscape created by the eruptions of 1730-1736.

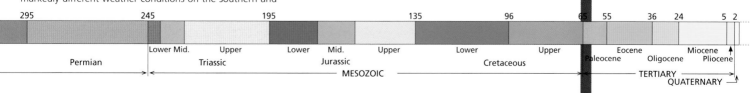

stone walls around each hole shelter the vines from the violent scorching winds. Built of pale volcanic stone, they stand in vivid contrast to the black ash, adding to the strange allure of one of the most fascinating viticultural landscapes in the world. Nature's reward for these Herculean efforts is the volcanic ash itself. Known locally as *picón*, it has hygroscopic properties, meaning that it readily absorbs moisture from the air and returns it to the soil in which the vines are rooted–a priceless advantage in a region with a mean annual rainfall of just five inches.

This type of planting is used more or less throughout the island of Lanzarote, although mainly in the central Geria Valley. At the northern end of the valley, where the land was covered by basalt lava flows, the wine growers first had to break through a tough basalt carapace more than three feet thick before planting the vines, usually one at a time, in the soft earth beneath. Some of this rootstock can be up to several feet in diameter.

These remarkable growing environments produce a variety of wines, with the emphasis on intensely aromatic, characterful white wines, based on the Malvoisie. This variety, together with the Muscatel, also yields the regional dessert wines.

Tenerife

The bulk of Canary Island wine comes from the island of Tenerife where the method of viniculture is more classical but still unusual. The island is formed from a single volcano (Mount Teide) with markedly different weather conditions on the southern and

northern sides. The southern slopes are hot and arid (Abona and Gümar Valley), while the northern flanks remain humid and temperate all year-round (in the appellation areas of Tacoronte-Acentejo, Orotava Valley, and Ycodén-Daute-Isora). The vines, mixed with other crops such as bananas, cover large areas of these steep slopes, and extend upwards to heights of more than 3,000 feet, planted in soils derived from the weathering of lava. The grape varieties are the Listán Blanco, Listán Negro, and Negramoll.

The unique vineyards of Lanzarote owe their survival to the hygroscopic properties of picón (volcanic ash).

295	245	195	135	96	65	55	36	24	5 2

Permian	Lower	Mid.	Upper	Lower	Mid.	Upper	Lower	Upper	Paleocene	Eocene	Oligocene	Miocene	Pliocene
	Triassic			Jurassic			Cretaceous						

MESOZOIC

TERTIARY

QUATERNARY

Brazil: The Serra Gaucha

There are vineyards throughout eastern Brazil, essentially concentrated in Rio Grande do Sul, Brazil's southernmost state. Bordering Uruguay to the south and Argentina to the west, this state accounts for 90 per cent of annual Brazilian wine production (three million hectoliters a year).

The capital of the State of Rio Grande do Sul, Porto Alegre, is located on the banks of a large coastal lake called the Laguna dos Patos, where there is no trace of vines. To find them, drive northwards for around an hour and a half, up to the Serra Gaucha Range. Halfway through the journey, the road starts to climb steeply through a lush, sheer-sided valley cascading with waterfalls. Eventually you come to the small town of Garibaldi, perched on an immense plateau that stretches as far as the eye can see. This is the Serra Gaucha Range, part of the Planalto des Araucarias in the northeast of Rio Grande do Sul.

The Serra Gaucha Range

The plateau is formed from successive layers of volcanic rock that were deposited over a very long period, extending from the Triassic to the Lower Cretaceous. More than 3,000 feet deep, these deposits essentially consist of basic igneous rocks including basalt*, andesites*, and volcanic and sedimentary breccia*, with acidic igneous rocks at the summit.

From the end of the Mesozoic, these very hard rocks came

Cross-section of the volcanic soils of the Vale dos Vinhedos.

The Vale dos Vinhedos: the vineyards of the Rio Leopoldina, which rises at the foot of the city of Bento Goncalves, are punctuated by rows of plane trees around the edges of each vineyard plot.

3 800	2 500	Millions of years		540	500		435	410		360	
Archaen		Proterozoic		Cambrian	Orodovician		Silurian	Devonian			Carbonifer
	PRECAMBRIAN							PALEOZOIC			

The growth of viticulture in this very particular environment goes back to the late nineteenth century, when Italian settlers from northern Italy pioneered the plantings of the American varieties Concord, Jacquez, Clinton, and especially the Isabelle. These varieties still account for nearly four-fifths of the total area under vine, although an increasing number of European vines have been introduced since the 1980s for the production of both still and sparkling wines. Plantings in the Serra Gaucha now feature not only the American varieties but also, for white wines, the Chardonnay, Italian Riesling, Sémillon, and Sauvignon Blanc; and for red wines, the Cabernet Sauvignon, Merlot, Pinot Noir, and Gamay Noir.

The vineyards of the Casa Valduga, in the Vale dos Vinhedos.

The Vale dos Vinhedos

This is without question the most dynamic part of the region. The aptly named Vale dos Vinhedos (Valley of the Vineyards) starts in the city of Bento-Goncalves (the heartland of Brazil's wine-producing region) and opens in a north westerly direction toward the Rio das Antas. The altitude ranges from 654 feet to more than 2,000 feet around Bento- Goncalves itself, on the plateau. The soils are volcanic in origin and vary from stony and superficial to deep and fertile. Grape vines grow in monoculture, either on steep slopes or more or less flat ledges. Because it is so wet here, most of the vines are grown on pergolas to keep the clusters off the ground, with the supporting wires attached to plane trees. In summer, the trees' dark green leaves contrast vividly with the light green of the vine, forming leafy green balls around the edges of each plot that give a remarkable rhythm to the landscape.

The wines made from European varieties, especially red wines, are of excellent quality and range from single cultivars to an increasing number of blends. Those based on the Cabernet Sauvignon and the Merlot, blended with the Cabernet Franc or the Tannat, are well built, with scents of red berries and the delicate spiciness of vanilla.

under attack from erosion* that wore away the softer outer layers. This left a sheer escarpment, marking the end of the plateau on the Atlantic side and overlooking the coastal plain of Santa Catarina from a height of more than 3,000 feet. The volcanic layers became deeply scoured by river channels. The region's main hydrographic basin is formed by the Rio das Antas that rises in São Francisco de Paula, east of the plateau, and then flows through an increasingly deep and steep-sided, V-shaped valley, forming endless meanders that create a landscape of spectacular beauty. It skirts northward, then westward, of the Serra Gaucha, where it meets up with the Rio Carriero and becomes the Rio Taquari. To the southeast, the Serra Gaucha is bordered by the Rio Cai that also flows in a deeply scoured channel. Bordering the range to the south is the great escarpment that towers over the plain of Porto Alegre. Gradually the flat volcanic layers forming the plateau were carved by erosion into a series of ledges that, from a distance, resembles a huge staircase, with towns at the top (Bento Goncalves and Garibaldi) and crops – especially vines – in the valleys on all sides.

The climate at the top of the plateau, at an altitude of 2,289-2,616 feet, is moderately hot with annual temperatures in the range of 60-64° F and abundant rainfall (approximately 66 inches a year, the maximum for grape vines). Cold, damp polar air masses from the south Atlantic in winter and can bring heavy snowfalls.

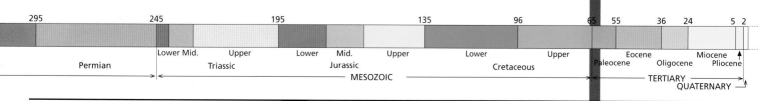

295		245		195		135		96	65	55	36	24	5 2
		Lower	Mid.	Upper	Lower	Mid.	Upper	Lower	Upper		Eocene		Miocene
Permian			Triassic			Jurassic			Cretaceous	Paleocene		Oligocene	Pliocene
						MESOZOIC						TERTIARY	
													QUATERNARY

A

Aalenian

Stage in the Middle Jurassic.

Accident

A surface irregularity in a natural formation.

Acidic

Of igneous rock having a silica content (SiO2) of more than two-thirds.

Albarizas

Characteristic soils of the Jerez region in Andalusia, derived from white Oligocene marls formed from the sedimentation of marine diatoms and radiolaria.

Albian

Stage in the Lower Cretaceous. The Maury terroir in the Roussillon is established on Albian schists.

Alkaline

Quality of potassium-rich (K) and sodium-rich (Na) environments.

Alluvia

Fine-grained sediment consisting of pebbles, gravel, sand, and silt, deposited by flowing water along the borders of river beds.

Alluvial fan

Accumulation of material stripped away from valley slopes and deposited by fast-flowing mountain rivers when they reach flatter land.

Alluvial plain

Flat zone between two massifs, consisting of an accumulation of stony, ancient alluvia or fine-grained recent alluvia.

Alveoline

Foraminiferal mollusk.

Amphibole

Dark-colored mineral of the silicate group, found in eruptive and metamorphic rock.

Amphibolite

Metamorphic rock consisting of amphiboles (silicates with two cleavage plains) and plagioclases (calcium-and sodium-rich feldspar), originating in zones of marked metamorphic activity.

Ancient Basement

Terrain that was violently folded, metamorphosed, and intruded by granite, then eroded and reduced to a peneplain in the course of a complete orogenic cycle. The ancient basement may or may not be overlain with sediments.

Andalusian Depression

Guadalquivir Basin in southern Spain.

Andesite

Effusive magmatic rock. Basic andesites form flows of rock. Thicker, acidic andesites form needles.

Anticline

A series of geological layers folded to form a convex arch in which the oldest rocks comprise the core. Antonym: syncline.

Aptian

Stage in the Lower Cretaceous. The Maury terroir in the Roussillon is surrounded by Aptian Urgonian limestones.

Aquitanian

Stage in the Lower Miocene.

Arène

Coarse sand derived from surface weathering of granite or gneiss.

Argovian

Name given to an Oxfordian substage (Upper Jurassic). Argovian marls are common on the Côte de Beaune.

Arkose

Sandstone consisting of grains of feldspar and quartz cemented by a mixture of quartz and clay minerals.

Astarte

Bivalve mollusc typically found in cold seas.

Astéries

Fragments of fossilized, star-shaped echinoderms (stelleroidea) found in *Calcaire à Astéries*.

Asthenosphere

Thin, semi-fluid layer of partially molten rock underlying the outer rigid lithosphere. The tectonic plates of Earth's crust float on the asthenosphere.

Aubuis

Stony clay limestone soils formed on the Turonian, characteristic of the Touraine region.

B

Bajocian

Mid-Jurassic stage. Hard Bajocian limestones often form sills in Burgundy, shielding the Liassic marls that support the vineyards (Montagny, Hautes Côtes de Beaune).

Barremien

Stage in the Lower Cretaceous. The vineyards on the right bank of the Rhône Valley are surrounded by hard Barremian limetones.

Barro

Soils similar to the *albarizas* in Jerez but sandier and less calcium-rich.

Bartonian

Stage in the Middle Miocene. The vineyards in the Minervois region are partly supported by the Bartonian.

Basalt

Effusive, extrusive igneous rock that periodically erupts along mid-oceanic ridges; flood basalt flows cool to form the very hard, compact horizontal sheets of rock now found in highland regions.

Basic

Of igneous rock with low silica (SiO2) content.

Bathonian

Stage in the Middle Jurassic. The strata package of the Côte de Nuits in Burgundy includes three Bathonian formations: Premeaux limestone, White oolite, and Comblanchien limestone.

Bauxite

Brightly colored (yellow or red) sedimentary rock, rich in aluminum hydroxide.

Berriasian

Stage in the Lower Cretaceous.

Bioclastic

Quality of sedimentary rock derived from shell fragments or similar organic remains.

Blocks

The terrain located on either side of a fault.

Boulbènes

Gascon term for alluvial deposits that have accumulated by solifluction on slopes and terraces.

Breccia

Rock consisting of more than 50 per cent angular rock fragments embedded in a finer matrix. Breccia may be of sedimentary or volcanic origin.

Brioverian

Final stage in the Precambrian. Brioverian schists are among the rocks supporting the vineyards of the Coteaux de Layon.

Brown soil

Soil with a brownish top layer due to the presence of clay and iron, and a lighter-colored, carbonate-rich substrate.

Buntsandstein

Another name for the Lower Triassic.

Burdigalian

Stage in the Lower Miocene.

C

Cadomian (cycle)

Orogenic cycle immediately preceding the start of the Cambrian (570 million years ago).

Caillotte

Pebbly, rendzina-type soils formed from Mid- and Lower Kimmeridgian limestones Synonym: *cris*.

Calcic

Containing or relating to lime or calcium.

Calcrete

Limestone crust that forms in soils due to the migration of calcium carbonate in percolating groundwater.

Callovian

Stage in the Middle Jurassic.

Campanian

Stage in the Upper Cretaceous. The excellent terroirs of the Champagne region are formed from Campanian chalk.

Carbonatite

Igneous rock composed mainly of carbonate crystals (calcite and/or dolomite).

Cargneules (or rauhwackes)

Sedimentary limestone rock with honeycomb aspect due to dissolution of certain elements of their initial composition.

Causse

Jurassic limestone plateau in the south and southwestern Massif Central of France.

Cenomanian

Stage in the Upper Cretaceous. The soils of the Cenomanian, the Sénonian, and the Turonian support the vineyards of the Touraine area.

Cenozoic

The most recent geological era, comprising the Tertiary and Quaternary periods.

Chalk

Soft, fine-grained white sedimentary rock deposited in shallow waters, consisting of nearly pure calcium carbonate and containing microscopic fossil fragments of unicellular marine organisms called coccoliths.

Chattian

Stage in the Oligocene. The vineyards of the Côte region in Switzerland, between Geneva and Lausanne, are established on Chattian molasse.

Cingles

The meander lobes of the Lot River, in the Cahors region of France.

Clay

Mineral consisting of superimposed laminae of hydrated aluminum silicate, essentially derived from weathered feldspar. Types of clay include montmorillonite or "swelling clays", so called because they expand after absorbing water.

Climax

The most intense phase of a geological process (such as uplift, erosion).

Cluse

Narrow valleys cut at right angles across folded ridges of relatively hard strata.

Colluvia

A mixture of rock fragments of variable coarseness moved short distances from their source and deposited at the bases of cliffs or hills.

Comblanchien limestone

Calcareous marble quarried at the commune of Comblanchien, in Burgundy on the Côte des Pierres, the rockiest part of the Côte-d'Or, located between the Côte de Nuits and the Côte de Beaune.

Conglomerate

Any coarse-grained sedimentary rock consisting of at least 50 percent rock fragments embedded in a finer matrix. Angular rock fragments are called breccia and rounded fragments are known as puddingstone.

Coomb

Narrow basin cutting across an anticlinal fold.

Corridor

Narrow valley between two massifs (e.g. the Rhône Corridor).

Côte

Long, steep hill between two more or less flat areas.

Coteau

Hillside.

Crais

Pebbly soils on *Calcaire à Entroques* (crinoidal limestone) in Gevrey-Chambertin in Burgundy.

Craton

Vast, stable part of the earth's crust or lithosphere that has not been deformed by orogeny.

Cris

Synonym for the *caillottes* of the Sancerre region.

Croupe

• Rounded hill

• French name for the gravel mounds of the Bordeaux region: strips of Quaternary terrace that have been dissected by small side streams and worn away along the edges by erosion.

Crust (Earth's)

The solid outer shell of the Earth, with an average thickness of 30-35 kilometers (18-21 miles) in continental regions, and 5 kilometers (3 miles) beneath the oceans. The continental crust contains every type of geological formation.

Crust (oceanic)

Part of the earth's crust located beneath the ocean. Oceanic crust is formed from effusions of basaltic magma at seafloor spreading centers (also known as mid-ocean ridges) where tectonic plates that border the ocean diverge. This magma then solidifies into new oceanic crust and is overlain by sediments around the margins of continents.

Crystalline

Quality of rocks formed from the crystallization of magma, essentially igneous (e.g. granite) and metamorphic rocks.

Cuesta

Long, low ridge formed by the differential erosion of alternating hard and soft sedimentary strata, with a steep scarp of hard strata jutting out over the softer underlying strata.

D

Dacite

Igneous, effusive rock that is highly viscous in magma form

Decalcification

In calcareous, loss of calcite rock as it is dissolved by water.

Delamarian Orogeny

The period of orogeny that affected Australia some 480-500 million years ago in the Ordovician, leading to the creation of the Mount Lofty Ranges, which are now home to some of the best vineyard sites in Australia (e.g. the Barossa Valley).

Deposit

An accumulation of fine and coarse sediments on a foundation. The terms marine, lacustrine, or continental refer to sediments that were originally deposited in seas, lakes, or on land, respectively.

Depression

Sunken area on the earth's surface.

Detrital

Condition of rocks formed from an accumulation of mineral debris or the skeletons of decayed organisms.

Dogger

Another name for the Middle Jurassic

Dolomite

Sedimentary rock consisting of at least 50 per cent calcium carbonate and magnesium.

Dolomitic (rock)

Limestone sedimentary rock in which calcite has been replaced by dolomite. Gave its name to the Dolomite Mountains in the Trentino-Alto-Aldige, Italy.

Drainage

Natural or built system that disposes of surface water.

E

Éboulis

Accumulation of usually angular stones as scree at the foot of hills. Very old *éboulis* can provide an excellent growing medium for vines.

Emerged

Of terrain that is no longer covered by water.

Entroques

Crinoidal rock debris formed from the delicate, feathery arms of echinoderms of the class *Crinoidea*, found in the Bajocian limestones of Burgundy.

Erosion

The wearing away of rock by physical or chemical processes.

Exogyra virgula

Bivalve, comma-shaped shellfish, rather like present-day oysters, characteristic of Mid- and Upper-Kimmeridgian marls.

F

Falun

Fossiliferous rock deposited by shallow seas.

Fault

A fracture in the Earth's crust resulting in the relative displacement and loss of continuity of the rocks on either side of it.

Fault displacement

The distance that any point on one side of a fault plane (within the same stratum) has moved in relation to a corresponding point on the opposite side.

Fault zone

Area of intense faulting and fracturing.

Fill

Accumulation of detrital or marine or lacustrine sediments in sunken areas of land.

Flint

Concretion of siliceous nodules typically occurring in limestone, especially chalk, formed by the precipitation of silica in semi-molten rock. Very hard, fracturing into sharp, erosion-resistant splinters. Produces very stony soils highly favorable to viticulture.

Folding

Deformation of a series of sedimentary layers due to flexion or torsion.

Fracture

Crack or fissure in geological formations.

G

Gabbro

Dark, coarse-grained, basic plutonic igneous rock, similar in structure to basalt.

Galestro

Tertiary schistose clay.

Gangue (or Deads)

Waste material such as stones, pebbles, and minerals that cement together the main constituents of rock.

Geological Epoch

A division of a geological period. The Jurassic is divided into three epochs: the Liassic, Dogger and Malm.

Geological Era

A major division of geological time, divided into several periods The three geological eras are: the Paleozoic (lasting 300 million years); the Mesozoic (lasting 180 million years); and the Cenozoic (lasting 70 million years). The latter is divided into the Tertiary period (lasting 69 million years) and the Quaternary (lasting one million years). The earliest geological era is the Precambrian that lasted about 4,000 million years, before the Cambrian period.

Geological Period

A division of a geological era. The Mesozoic Era is divided into three periods, the Triassic, Jurassic, and Cretaceous.

Geological Stage

A subdivision of a geological epoch. The Dogger (middle Jurassic period) is divided into the Aalinian, Bajocian, Bathonian, and Callovian stages.

Glacis

Slight incline made from a veneer of alluvial or colluvial deposits sometimes known as an alluvial glacis.

Gneiss

Fairly widespread metamorphic, coarse-grained rock consisting of quartz, feldspar, and mica.

Gondwana (or Gondwanaland)

Ancient super-continent of the late Paleozoic Era that in the course of the Mesozoic and early Cenozoic (Tertiary) comprised what are now Africa, South America, Australia, Antarctica, and the Indian subcontinent.

Gore

Pinkish granitic arenaceous soils found in the Beaujolais and Roanne regions.

Graben

Elongated, generally flat-bottomed topographic trough between two faults or two sets of faults (e.g. the Bresse graben on the Alsace plain).

Graben formation

Subsidence of the earth's crust between two faults

Granite

Light-colored, coarse-grained acidic plutonic igneous rock consisting of quartz, feldspar, and mica that forms the bedrock of most continents.

Graywacke

Sedimentary detrital rock containing rock debris, grains of quartz and feldspar with a clay matrix.

Gravel (*Graves*)

Pebbly, gravelly Quaternary alluvia characteristic of the Bordeaux region.

Gravel

Mixture of rock fragments with diameters in the range 2-30 millimeters (0.07-1.18 inches).

Grès bigarré

Sandstone of the Buntsandstein (corresponding to the Lower Triassic) with variegated color owing to the presence of iron oxide (red) and glauconite (green).

Grèze

Rock fragments of varying coarseness in a clay gangue, spread down the slope of a hill.

Gypsum

A relatively soft, colorless, translucent or milky mineral composed of crystals of hydrated calcium sulphate ($CaSO_4\ H_2O$).

H

Hauterivian

Stage in the Lower Cretaceous.

Helvetian

Stage in the Middle Miocene.

Hercynian Basement

Peneplain produced by the Hercynian Orogeny and razing of the Hercynian range, the surface being subsequently overlain by Mesozoic and Tertiary sediments.

Hercynian Range (or Massif)

Group of mountains formed in the Hercynian cycle that occurred in the Paleozoic Era from the Devonian to the Permian, and that came under long erosion at the end of the Paleozoic. There are significant outcrops of the Hercynian Range in Europe and North America (in the Appalachians).

Horst

Ridge of land, possibly consisting of a block of basement, overlain by sediments, that has been forced upward between two parallel faults.

Hydromorphous

Of soils exhibiting characteristics due to permanent or temporary excess groundwater on part or all of the slope. Hydromorphous soils are not suitable for viticulture.

I

Incursion (marine)

Sea inundation of an area that was previously above water, either due to subsidence of the newly flooded terrain or to an overall rise in sea level.

Inoceramus

Genus of bivalve found in the Jurassic and Cretaceous seas. Chalks containing this bivalve form the base of the Turonian beneath the white tuffeau of the Touraine area.

Inverted relief (formation)

Process of morphological transformation in which valleys become hills and vice versa through erosion. For instance, certain layers of basalt and Quaternary terraces that originated as valley-floor deposits are now positioned at the tops of hills, due to the erosion of the surrounding land.

K

Karst

Hard limestone plateau terrain where much of the calcium carbonate is leached away. Karstic scenery is characterized by dry valleys, subterranean drainage, and sinkholes.

Keuper

Stage in the Upper Triassic. One of the many formations that outcrop in the sub-Vosgian hills of Alsace.

Kimmeridgian

Stage in the Upper Jurassic, notably associated with the Chablis terroirs.

L

Lacustrine (limestones)

Limestone sediments deposited at the bottom of lakes.

Lagunal limestone

Limestone deposited at the bottom of lagoons.

Laurasia

Ancient supercontinent of the Mesozoic Era, corresponding to the northern part of Pangaea after it split from Gondwanaland and comprising what are now North America and Eurasia.

Lava

Molten rock that erupts from volcanoes. Lava can be fluid or viscous depending on its chemical composition (acidic or basic), its effusive temperature, and the environment (above ground or underwater).

Liassic

Another name for the Lower Jurassic. The white wines of Montagny come from Chardonnay vines planted in Liassic marls.

Limestone

Sedimentary rock consisting of at least 50 per cent calcium carbonate.
Limestone flagstones
Slab of hard calcite with well-defined edges.

Lithosphere

The rigid outer layer of the Earth, composed of the outer crust and the very top of the mantle and divided into large plates that drift across the asthenosphere.

Lœss

A light-colored, fine-grained accumulation of quartz, limestone, and clay particles, that have been deposited by the wind, especially the violent winds that marked the Quaternary glacial epochs.

Ludian

Stage in the Eocene.

Lutetian

Stage in the Middle Eocene. Origin of the detrital molasse in the north of the Limoux area.

M

Maastrichtian

Stage in the Upper Cretaceous. Origin of the red marls that appear in the south of the Limoux area.

Macigno

Oligocene millstone characteristic of the Tuscany area.

Magma

Molten rock within the ground at a temperature of 1112 °F. Magma may solidify deep inside the earth (to form plutonic rock), in shallow crust near the surface (to form granite) or on the earth's surface (to form volcanic rock).

Malm

Another name for the Upper Jurassic.

Marl

Fine-grained sedimentary rock consisting of around 50 per cent clay, characteristic of Mesozoic and Tertiary deposits. Marls frequently alternate with harder limestones.

Massive limestone

Thick layers of hard limestone, eg Urgonian limestone.

Mesogee

Synonym of Tethys.

Mesozoic

The middle geological era, after the Paleozoic.

Messinian

Stage in the Upper Miocene in which the Mediterranean Sea was cut off from the Atlantic Ocean and dried up.

Metamorphism

The mineralogical, chemical, and especially crystallographic changes in solid-state rock resulting from pressure and/or heat.

Metamorphosed

That which has undergone metamorphism.

Mica schists

Medium-grained metamorphic rock with foliate structure, rich in leaf-like layers of mica, that can be easily split into laminae. Consists of quartz, small amounts of feldspar, and other minerals depending on the type of rock.

Mill stone

Hard, compact, or porous rock composed of irregular masses of silica and limestone, used as a building stone in the Paris region.

Molasse

Soft, sometimes thick detrital sediments, ranging in composition from silt to sandstone or conglomerate, shed around uplifting mountains. The name molasse is thought to be derived from the sandstone used as millstones in Switzerland.

Monoclinal block

Block of stratified rock in which the strata are inclined at the same angle from the horizontal.

Montille

Strips of stony terraces surrounded by a zone of recent alluvia in the Touraine area.

Moraine

Mass of debris (stones, gravel, and sand) stripped from the sides of valleys by glaciers. Terminal moraines develop at the fronts of glaciers; lateral moraines form along the margins of glaciers.

Morgon

Arène derived from in situ rock weathering in the Beaujolais area.

Mother rock

Hard or soft rock that weathers superficially to produce the topsoil.

Mourrel

Small hillock found in the Minerve area, formed from a sandstone bank that resisted erosion surrounded by Bartonian molasse (Eocene).

Muschelkalk

Another name for the Middle Triassic. The marly soils of the great Marlenheim Grand Cru in Alsace are overlain by limestone scree of the Muschelkalk.

N

Namibian Orogeny

The period of orogeny that affected South Africa at the end of the Precambrian.

Nappe

A group of geological formations that has been thrust by orogenic displacement from its original foundation onto another, more or less distant foundation (e.g. the nappe in the eastern Corbières).

Neck

Solid block of lava from the opening of an extinct volcano, exposed after erosion of the surrounding rock.

O

Oceanic ridge

Any section of the range of mid-ocean submarine mountains that extend many thousands of miles. Sites of intense underwater volcanic activity, oceanic ridges generally feature an axial graben or rift at the diverging plate boundary from which basalt magma escapes.

Oolith

Tiny spherical grain (0.03-0.07 inches in diameter), having a central nucleus (quartz or fossilized debris) surrounded by cemented, concentric layers of calcium or ferruginous material.

Oolitic

That which contains ooliths.

Ophiolite

Basic, metamorphosed magma produced by oceanic rifts, seen today, for example, in the Alps.

Orogeny

Series of processes leading to the formation of mountain ranges (e.g. the Hercynian and Alpine Orogenies).

Ostrea acuminata

Bivalve mollusk belonging to the same family as oysters, characteristic of the marls that cap the Bajocian crinoidal limestone in Burgundy.

Outcrop

Part of a rock formation or mineral vein that appears at the surface of the Earth. In France, the Hercynian basement outcrops in Brittany, the Massif Central, and the Vosges.

Oxfordian

Stage in the Upper Jurassic. The hills of the Côte de Beaune in Burgundy are capped with Oxfordian limestone.

P

Paleozoic

The first geological era, after the Precambrian.

Pangaea

The ancient super-continent comprising all the areas of non-submerged land at the end of the Paleozoic, that then split into two to produce Laurasia and Gondwanaland.

Páramo

Spanish for plateau. The vast, hard limestone plateaus of the Pontian in Spain, carved by the rivers of Castilla-León.

Pebble

A smooth, rounded rock fragment, shaped by the action of water.
Pebbly sands
Sands interbedded with pebbles, often present in reshuffled fluvial alluvial deposits

Pelite

Fine-grained, detrital sedimentary rock, consisting of clay, quartz, and fossilized debris.

Peneplain
Flat or lightly undulating land surface produced in the final phase of the erosion of a mountain range.

Peperite
Pyroclastic limestone or marly rock containing peppercorn-like grains of basalt (smaller than 0.5 inches in diameter) that formed in volcanic chimneys as a result of phreatomagmatic explosions.

Perruche
Very stony soils on flinty clays of the Sénonian (Upper Cretaceous), found in the Touraine area on the upper slopes overlooking the Loire and some of its tributaries.

Phtanite
Dark sedimentary rock formed from microscopic, tightly packed, stratified quartz crystals; mainly found in Precambrian and Paleozoic terrain (such as the Coteau du Layon).

Phyllite
Argillaceous mineral of the phyllosilicate family (micas).

Piedmont
Area at the foot of a mountain range with a gentle to moderate slope, formed from colluvia and alluvia shed by surrounding hills.

Pierre dorée
Fawn-colored Aalenian crinoidal limestone, widely quarried as a building stone in the Beaujolais region.

Plate
Any of the rigid layers of the Earth's crust (lithosphere) floating over the viscous asthenosphere. Divergent plate boundaries (rift zones in which two tectonic plates are pulling apart) are associated with lava effusions from the tear in the oceanic crust. Convergent plate boundaries (where two plates are colliding) are associated with the formation of mountain ranges (Andes, Alps).

Plateau
A wide, mainly level area of elevated land with sharp edges.

Pontian
Another name for the Messinian.

Porphyry
Hard, igneous reddish-purple rock consisting of large crystals of feldspar in a finer groundmass.

Portlandian
Stage in the Upper Jurassic. The Chablis vineyards are planted on Portlandian formations.

Portlandian ledge
Ledge of hard Portlandian limestone that forms the classic cap rock along the southeastern part of the Paris Basin.

Premeaux limestone
Limestone from the commune of Premeaux on the Côte d'Or in Burgundy, quarried as a building stone.

Pudding stone
Sedimentary, detrital rock mainly consisting of rounded pebbles within a finer matrix.

Pyroclastic
Debris formed from igneous rock ejected during a volcanic eruption.

Q

Quartz
Crystalline form of silica (SiO_2).

R

Rauracian
A division of the Oxfordian.

Red sands
Sands colored red by iron or aluminum oxides.

Relaxation Phase
Release of the pressures of orogenic compression in the course of mountain building.

Relief
Variation in altitude in an area; degree of difference between highest and lowest level, e.g. a region of low relief.

Remnant hills
Isolated hills, often capped by resistant rock, remaining after erosion has worn away surrounding rocks.

Rendzina
Soil developed from limestone mother rock, characterized by a topsoil rich in organic matter and carbonates, overlaying a pebbly limestone substrate.

Restanque
Fairly narrow terraces built by winegrowers in Provence to make planting easier on steep slopes.

Rhyolite
Light-colored, effusive igneous rock producing domes or short, thick coulées of viscous lava.

Rift
A fault produced by tension on either side of a fault plane. The rifts in the center of the mid-ocean ridges are sites of intense volcanic activity.

Rudist
Bivalve found in reef corals in the warm seas of the Jurassic and Cretaceous.

Ruedas
Red, sandy-limestone soils characteristic of the Montilla-Moriles area of Andalusia.

Ruffe
Fine red Permian sandstone found in the Lodève region (Languedoc), producing more or less stony sandy-argillaceous soils that are quite favorable to viticulture.

Rupelian
Stage in the Oligocene. The Swiss plateau is partly formed from Rupelian marine sediments.

S

Safres
Name of the sandy Miocene marls of the southern Rhône Valley.

Sand
Loose detrital sediment formed of quartz grains less than 0.7 inches in diameter.

Sandstone
Detrital sedimentary rock consisting of at least 85 per cent sand grains consolidated with such materials as quartz. Color and hardness vary depending on the cement.

Sannoisian
Stage in the Oligocene.

Sarmatian
Central European term for the Late Middle Miocene.

Schist
Any metamorphic rock that can be easily split into thin layers, because its micaceous minerals are aligned in thin parallel bands. Schists may be of sedimentary origin (such as pelites) or produced by the metamorphism of rocks due to the action of plate tectonics. Quarried as slate.

Schiste lustré
Calcium-rich metamorphic rock blueschist originates in underwater volcanic eruptions (ophiolites) in the middle of the Tethys Sea. There are considerable sheets of blueschist in the Alps. Ophiolites were metamorphosed in the course of the Alpine uplift.

Scree – see *éboulis*

Secondary chain or range (of mountains)
Group of similar mountains bounded by valleys within a larger range.

Sedimentary Basin
Depression in the Earth's surface exhibiting tectonic subsidence and consequent infilling by continental or marine sedimentation.

Sedimentary covering
The volcanic or sedimentary material overlaying basement rock that has undergone a complete orogenic cycle (folding and uplift followed by leveling from long periods of erosion).

Sedimentary rock
More or less hard rock formed from material that has been deposited by seas or on dry land (by gravity or water).

Sediment Deposition
The accumulation of deposits on a basement. There are three major types of deposition: marine, transitional (lagoons, deltas, etc.), and continental (fluvial, lacustrine, windblown).

Sénonian
Stage in the Upper Cretaceous. The *perruches* soils of the Touraine area are derived from flinty clays of the Sénonian.

Séquanien
Name for a division of the Oxfordian.

Serpentine
Soft green rock with a snake-like lustre, produced by metamorphism of basic igneous rock.

Sidérolithique
Formation consisting of red clays rich in ferruginous nodules, derived from weathered limestone rocks in particular. Characteristic of the eastern Aquitaine Basin.

Silesian
Name of the coal formation period of the Carboniferous.

Silt
Very fine-grained detrital and often argillaceous sediment, originating in continental, fluvial, or wind-blown deposits.

Sinemurian
Stage in the Lower Jurassic. A few of the Hautes Côtes de Beaune vineyards in Burgundy are supported by Sinemurian limestones.

Slate
Brown to bluish-gray fine-grained metamorphic rock with a layered (foliate) texture, extracted from slaty schist.

Slaty schist
Metamorphic rock resulting from the lowest temperature and pressure conditions.

Soil
The top layer of the land surface produced by weathering of subjacent mother rock and, more recently, human effort.

Solifluction
Slow downhill movement of soil saturated with meltwater. The term is associated with the sliding of a terrain as a coherent mass.

Spilite
Dark green igneous rock similar to basalt.

Stampian
Stage in the Oligocene. The *Calcaire à Astéries* that characterizes so many of the Bordeaux terroirs is a Stampian formation.

Subduction
The movement of the oceanic crust beneath the continental crust as one tectonic plate slides under another, often resulting in oceanic rifts along the point of subduction and intense volcanic activity along the continental border. The Andes were formed by subduction.

Sublithographique (limestone)
Fine-grained, homogeneous limestone, less compact than lithographic limestone which is the type formerly used in printing.

Subsidence
Gradual sinking of terrain under the weight of accumulated sediments.

Syncline
A concave fold involving a series of geological layers, with the oldest rocks on the outside of the fold. Antonym: anticline.

T

Tectonic

Structural deformation of a group of geological strata following their formation or deposition (folding, rupture, overthrust). Current theories of plate tectonics define continents as being part of tectonic plates that are floating on the asthenosphere.

Terra rossa

Soil formed in a marine environment, composed of red clay and the residue of dissolved limestone.

Terrace

Shelf-like surface bordering a body of water but slightly higher, formed by an earlier phase of pebbly deposition, manmade shelf-like surfaces on steep slopes, designed to facilitate planting on higher ground.

Terres blanches

The steep terrain on Kimmeridgian marls found in the Sancerre region.

Tethys

Vast sea separating Eurasia from Africa before the Alpine collision (250 million years ago). All that remains of this sea today is the Mediterranean.

Thanetian

Stage in the Paleocene. The vineyards of the Minervois are supported by Thanetian formations.

Toarcian

Subdivision of the Lower Jurassic.

Tortonian

Stage in the Upper Miocene. There are outcrops of the Tortonian in the eastern section of the Barolo area, in the Italian Piedmont region.

Trachyte

Highly viscous, effusive magmatic rock, rich in feldspar but free of quartz, forming volcanic plugs.

Tran

Water-resistant layer of clay and iron deep within the soils of Pécharmant in the Bergerac area.

Tuff

Rock formed by small rock fragments ejected from a volcano and bound together by water.

Tufa

Limestone concretion that forms around certain water sources through calcium precipitation.

Tuffeau

Chalky, more or less sandstone-rich limestone of the Turonian (Upper Cretaceous) found in the Touraine. Extensively quarried as a local building stone, notably for the celebrated Loire châteaux. Yellow tuffeau is softer and was excavated to make dwellings along the valley of the Loire and its tributaries.

Turonian

Stage in the Upper Cretaceous, corresponding to the geological layer of tuffeau that has been dug out to make the cellars of the Touraine area.

Turritelle

Mollusk-like gastropod with spiral shell, first appearing at the end of the Mesozoic Era.

U

Uplift

Abrupt or gradual uplift of a more or less vast landform under pressure, e.g. when two tectonic plates collide.

Urgonian (limestone)

Very hard, dazzlingly white and massive limestone, deposited in the Lower Cretaceous around the Vocontian Trough.

V

Valanginian

Stage in the Lower Cretaceous.

Veneer

Shallow deposits that are in discordant contact with the foundation, e.g. loess veneers.

Villafranchian

Intermediate stage between the Tertiary and Quaternary periods. Stony sheets of Villafranchian deposits are one of the major viticultural supports of the Rhône Valley.

Villafranchian nappe

Vast expanse of ancient alluvia, rich in pebbles washed down by rivers and streams from the Alps and the Pyrenees at the junction of the Tertiary and Quaternary periods, e.g. on the plateau of Châteauneuf-du-Pape.

Volcano

Mountain formed from magma ejected from a vent in the Earth's crust. Depending on the chemical composition of the magma ejected, volcanoes can form plugs, domes, flared cones, or virtually horizontal lava flows.

Volcanic ash

Fine particles of lava thrown out by an erupting volcano.

Volcanic rock

Effusive magmatic rock that spreads over the Earth's surface or beneath the oceans (rifts).

Volcanism

Those processes that result collectively in the formation of volcanoes and their products.

W

Weathering

The disintegration of rock into smaller fragments of the same mineral composition by processes such as frost action or water solubility.

Wind-blown

Relating to the wind; produced or formed by the wind.

Y

Ypresian

Stage of the Lower Eocene

M = maps, geological cross-sections and block diagrams, C = captions. • Page numbers in bold type indicate the main references.

Photographic Credits

EDITOR
Catherine Montalbetti

ASSISTANT EDITOR
Christine Cuperly

ENGLISH TRANSLATION BY
Florence Brutton

SECRETARIAL ASSISTANT
Anne Le Meur

PROOF READING
Élisabeth Bonvarlet

MAP DESIGN
Block diagrams and cross-sections:
Hachette Education Cartographic Department
P. 27: Fabrice Le Goff.

GRAPHIC DESIGN
Graph'm / François Huertas

PRODUCTION
Graph'm

Printed in China
by Graficas Estella
Copyright registration: 14307-11-2001
Edition: OF 08968
ISBN 201 236 504 3
23.51.6504-7/01

Cover: Châteuneuf-du-Pape (Scope/J. Guillard); Douro Valley, Pinhão region, Portugal (Scope/J. Guillard); Aloxe-Corton (Scope/J. Guillard); Sancerre Clos de La Poussie (Scope/M. Plassart).

Jacques Fanet : 5 (right), 17, 18, 20, 21, 49, 50-51, 73, 108, 111, 112, 113, 117, 122 (center), 130 (top), 140, 141, 142, 143, 149, 166, 167, 174, 193 (x 2), 200, 218, 222, 223 (bottom), 226, 227, 228, 229.

Photothèque Hachette Livre: 14, 15 (bottom).

C. Sarramon: 58, 87, 119, 120, 121, 125, 183 (bottom).

SCOPE
- ***J.-L. Barde:*** 48 (x 2), 52, 55, 56, 57, 59, 61, 67, 78 (top), 79 (top), 86, 88, 89, 94 (top x 2), 95 (bottom), 98, 99, 103, 104, 105, 106, 115, 116, 118, 123 (bottom), 124, 130, 135 (bottom), 152-153, 157, 158, 163, 173, 187, 189, 190, 191, 209, 210, 215, 221.
- ***P. Beuzen :*** 123, 216, 217 (top).
- ***P. Blondel:*** 80.
- ***C. Cheadle:*** 131.
- ***P. Galeron:*** 97 (top).
- ***M. Gotin:*** 178, 225.
- ***J. Guillard:*** 4, 5 (left), 10, 22, 24, 26, 27 (x 2), 28 (x 2), 29, 30 (x 2), 31, 32 (x 2), 33, 34 (x 3), 35, 37, 38, 39, 44 (x 2), 46, 47 (x 2), 60, 62, 64, 65, 66 (x 2), 68, 72, 74, 75 (x 2), 77, 78 (bottom), 83, 100, 109, 118 (top),122 (top), 126, 127, 128 (x 2), 129, 131 (bottom), 132-133, 134, 135 (top), 137, 138, 139, 146, 162, 168, 169, 171, 172 (x 2), 179, 180 (x 2), 181, 183 (top), 184, 185, 194, 196, 197, 198, 199, 213, 214, 223 (top).
- ***M. Guillard:*** 71, 76, 82, 84, 90, 91, 92, 94 (bottom), 95 (top), 150, 152 (top), 154 (x 2), 159, 160, 164, 165, 212.
- ***Kactus/Hugues:*** 155, 201.
- ***S. Matthews:*** 85, 97 (bottom), 144, 147, 203, 204, 205, 207, 222.
- ***M. Plassart:*** 15, 176.
- ***N. Servan:*** 79 (bottom).
- ***D. Sudres:*** 83.

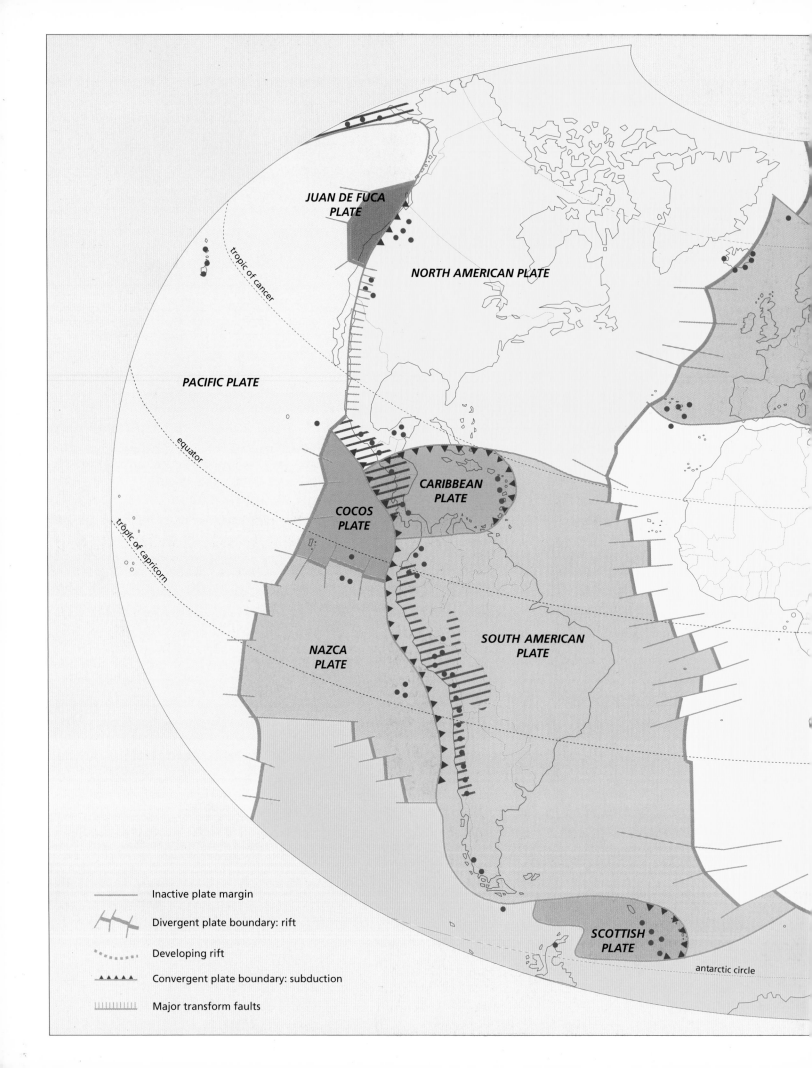

JUAN DE FUCA
PLATE

NORTH AMERICAN PLATE

PACIFIC PLATE

tropic of cancer

equator

tropic of capricorn

COCOS
PLATE

CARIBBEAN
PLATE

NAZCA
PLATE

SOUTH AMERICAN
PLATE

SCOTTISH
PLATE

antarctic circle

Inactive plate margin

Divergent plate boundary: rift

Developing rift

Convergent plate boundary: subduction

Major transform faults